景观中国
LANDSCAPE CHINA
3

佳图文化 编

中国林业出版社
China Forestry Publishing House

图书在版编目（CIP）数据

景观中国 . 3 / 佳图文化主编 . -- 北京：中国林业出版社，2016.9

ISBN 978-7-5038-8562-4

Ⅰ . ①景… Ⅱ . ①佳… Ⅲ . ①景观设计 – 案例 – 中国 Ⅳ . ① TU986.2

中国版本图书馆 CIP 数据核字 (2016) 第 129239 号

中国林业出版社·建筑家居出版分社
责任编辑：李 顺　李 辰
出版咨询：（010）83143569

出　版：中国林业出版社（100009 北京西城区德内大街刘海胡同 7 号）
网　站：http://lycb.forestry.gov.cn/
印　刷：广州中天彩色印刷有限公司
发　行：中国林业出版社
电　话：（010）83143500
版　次：2016 年 9 月第 1 版
印　次：2016 年 9 月第 1 次
开　本：889mm×1194mm 1 / 16
印　张：22.5
字　数：200 千字
定　价：320.00 元

Preface 前言

As the sequel of *Beautiful China I*, *Beautiful China II* selected the representative landscape for housing and public business in China and continued to professionally analyze these "beautiful" cases in detail from multiple perspectives. Content layout starts from the aspects of the keywords, features, and design concept, providing a large number of professional technical drawings, with rich and detailed information. Our effort is to give landscape designers and practitioners new vision and inspiration, create more beautiful landscape, and achieve the Chinese dream of constructing beautiful China and realizing the Chinese nation's sustainable development in landscape design.

《美丽中国Ⅱ》是《美丽中国Ⅰ》的延续之作，精选了国内具有代表性的住宅景观以及公共商业景观，继续从多角度详细且专业地分析了这些"美丽"的景观案例。内容编排上，分别从景观案例的关键点、亮点、设计理念等方面入手，配合大量的专业技术图纸，资料丰富而详实。我们的努力是为了给予设计师及景观从业者新的视觉、新的灵感，以创作出更多更美的景观，在景观设计上共同实现建设美丽中国、实现中华民族永续发展的中国梦。

CONTENTS

PARK LANDSCAPE
公园景观

002　Sports Park Landscape Renovation and Expansion, Dongcheng District, Dongguan　东莞东城区体育公园改扩建景观工程

008　Changsha Meixi Lake Peach Blossom Mountain Park　长沙梅溪湖桃花岭山体公园

016　Zhangzhou Bihu Ecological Park　漳州碧湖市民生态公园

COMMERCIAL LANDSCAPE
商业景观

036　Wangjing SOHO　望京 SOHO

044　Greenland International Expo City, Nanchang　南昌绿地博览城

058　Landscape Design for La Vendome, Pazhou, Guangzhou　广州市琶洲南丰汇景观设计

064　Changshu Merchants Culture Commercial Plaza　常熟招商文化商业广场

074　K Wah Crowne Plaza　嘉华皇冠假日酒店

080　Mayland Lake Hot Spring Hotel　美林湖温泉大酒店

090　Xinhua City International Plaza Commercial Area Landscape, Tai'an　泰安新华城国际广场商业区景观

098　Chongqing Vanke Xijiu Plaza　重庆万科西九广场

104　Dongshenghui　东升汇

114　Vanke Northern Dream Town, Guangzhou　万科广州北部万科城

128　Powerlong Plaza, Suqian, Jiangsu　江苏宿迁宝龙广场

138　Haishang Bay Holiday Apartments, Financial Street of Xunliao Bay, Huizhou　金融街惠州巽寮湾海尚湾畔度假公寓

目录

152　Vanke Tsingtao Pearl　万科青岛小镇

160　Tianjin Hedong Wanda Plaza　天津河东万达中心

168　Thaihot City Plaza, Wusibei, Fuzhou　福州五四北泰禾城市广场

178　Lingshui Sandalwood Resort Phase Ⅱ　陵水红磡香水湾度假酒店二期

188　Dream Island Sales Center, Nanjing　南京仁恒绿洲新岛销售中心

RIVERSIDE LANDSCAPE
滨江景观带

198　Landscape Design for the Bund Source 33 Park and Suzhou River Waterside Platform
　　　外滩源 33 公共绿地及苏州河亲水平台环境景观设计

206　Zhangjiagang Xiaocheng River Renovation　张家港小城河改造

220　Foshan Dongping New Town Riverfront Landscape Zone　佛山东平新城滨河景观带

238　West Peninsula of Zhongda Group, Fuyang　富阳中大西郊半岛

LANDSCAPE FOR CULTURAL & CIVIC BUILDINGS
文化、市政建筑景观

248　Dongying City Lv Opera Museum Square Landscape　东营市吕剧博物馆广场景观

256　Science and Technology Innovation Park of Zhejiang University, Luoyang　洛阳浙大科技创意园

Park Landscape

公园景观

Ecological Space
生态空间

Core Landscape
核心景观

Sustainability
可持续性

Park Landscape
公园景观

KEYWORDS 关键词

Rich Visual Effect
视觉丰富

Landscape Wall
特色景墙

Cultural Atmosphere
人文特质

Modernism Style
现代主义风格

Location: Dongcheng District, Dongguan, Guangdong
Client: Dongcheng District Office of Dongguan City
Landscape Design: Keymaster Consultant Co., Ltd.
Design Team: Z.Y. Cheung, Huang Wanjun, Huang Zhiping, Fu Wenfeng, Liu Qing
Land Area: 20,000 m²

项目地点：广东省东莞市东城区
业主：东莞市东城区办事处
景观设计：广州市科美都市景观规划有限公司
设计团队：张兆云、黄婉君、黄志平、符文峰、刘清
占地面积：20 000 m²

Sports Park Landscape Renovation and Expansion, Dongcheng District, Dongguan

东莞东城区体育公园改扩建景观工程

FEATURES 项目亮点

The sports park around the stadium is mainly designed with simple and elegant lines. Big trees, green lawns and stylish lamp poles are combined to create continuing spaces and provide special walking experience. Besides, there are also specially designed sculptures and landscape walls to form dynamic activity spaces, which highlight the cultural atmosphere of the sports park and create a new landmark for Dongcheng District.

整体设计以简洁大气的线条为主，利用大树、造型草坪与特色灯柱相结合，创造视觉丰富的人行体验和延续性空间，并结合个性雕塑及特色景墙，营造动感的活动空间，彰显该体育公园的人文特质，旨在打造为标示性的地区景观，成为东城区新兴的形象焦点。

Overview

The sports park around the stadium is mainly designed with simple and elegant lines. Big trees, green lawns and stylish lamp poles are combined to create continuing spaces and provide special walking experience. Besides, there are also specially designed sculptures and landscape walls to form dynamic activity spaces, which highlight the cultural atmosphere of the sports park and create a new landmark for Dongcheng District.

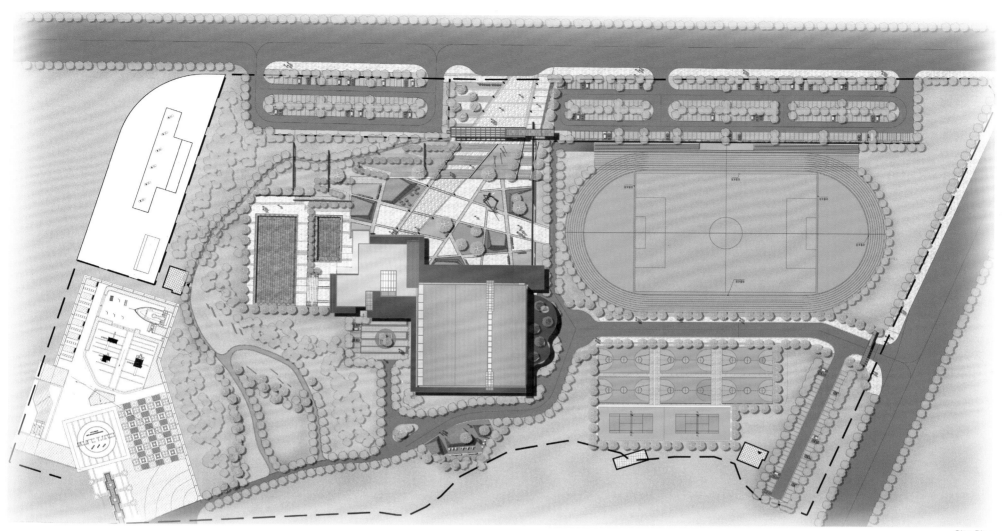

Site Plan
总平面图

项目概况

　　该项目为体育场馆外的公园景观设计，整体设计以简洁大气的线条为主，利用大树、造型草坪与特色灯柱相结合，创造视觉丰富的人行体验和延续性空间，并结合个性雕塑及特色景墙，营造动感的活动空间，彰显该体育公园的人文特质，旨在打造为标示性的地区景观，成为东城区新兴的形象焦点。

Design Goals

A stadium is not only an important venue for sports events but also a place for people to do exercise and relax themselves. What's more, it should have a public square for large-scale activities or performances.

Therefore, the designers have referred to some excellent designs for public squares and parks, for example, the Grand Canal Square in the U.S.A. with clean and elegant pavement as well as LED strips; the Olympic Sculpture Park in Seattle with symbolic sculptures and sports spirit; Shanghai Lotus Park with elegant and green vegetation and pavement; etc. and have got some important design elements.

The interaction between pavements, lighting, sculptures, plants and human beings will be the focuses of the design.

1. Dynamic oblique lines interweave on the site to bring vitality to the park together with the multi-level green spaces.

2. Large symbolic sculptures that embody the spirit of the Olympic Games and small interesting sculptures that interact with people have shown the "strength", "speed" and "competition" in sports.

3. Benches, walls, green vegetation, steps and ground pavement are designed with LED lights to create special atmosphere at night and gather people in the square even in the evening.

4. Different plants are well employed and organized to form neat lawns and grass slopes as well as linear ground covers and point-type lawns, providing the space with experience of large square, great lawn and large space.

设计目标

该项目作为一个市政体育场馆的外景观改造，场馆既要担当比赛举行的重要场地，同时平时也要兼备一个满足人们日常体育运动及休闲健康生活的功能要求，并具备一个公共广场举行大型活动或者表演的功能需求。

因此，设计师分析和总结了国内外的一些公共广场及公园的案例：如美国京航大运河广场（干净简洁的铺装、LED灯带的运用）；西雅图奥林匹克雕塑公园（标志性雕塑及体育精神的融入）；上海莲花公园（整洁的绿化种植及铺装感受）等，提取与总结了一些重要的设计元素。

铺装、灯光、雕塑、植物与人之间的互动性，将成为本次的设计重点。

1. 利用动感的斜线交织整个场地，并结合高低变化的趣味造型绿化，提升场地的活力感与趣味性。

2. 集中设置大型体现奥林匹克运动精神的标识雕塑及分散设置与人产生互动的趣味性雕塑小品，展现体育的力量感、速度感与竞技感。

3. 线性的座凳、墙体、造型绿化、台阶、地面铺装等方面设计LED灯带，利用灯光打造广场的夜景氛围，使广场在夜晚也能成为一个聚集人气的活动广场。

4. 在植物种植上，强调干净的草坪、草坡、线状整形地被及点状的大草坪种植为原则，塑造大广场、大草坪、大场景的空间感受。

Square Space

Although the park is not big, the design process has well shown the communication and cooperation between the design firm, the construction team, the sculpture firm and the client. The designers have kept the initiative during this process and tried to realize the design idea. As a public space, the design for it is different from that for residential communities. More attention should be paid to the square space and the circulation. The square space provides more opportunities for varied public activities such as kicking shuttlecock, square dance, large outdoor performance, etc., which will be sure to encourage mutual communications. Thus the sports park can function as a place for sports and relaxation, and at the same time to provide great potentials for future functions.

广场空间

该项目虽然不大，但在过程中却很好地体现了设计公司与施工队、雕塑公司及甲方的多层次、多方面的沟通与合作，使得设计师在设计上保留了较大的主动权，使得场地景观较大限度地体现了其初衷。而作为一个市政公共空间的设计，区别于平时的住宅小区的设计原则，更应注重广场空间及人流的组织，广场空间的设置为市民提供各种互动活动提供了更多的可能性，从而引发人展开各种交流及互动的行为产生，例如踢毽子、广场舞、大型户外表演等，思考的出发点从人的公共性及互动性展开，从而希望体育公园能承载它的使命，成为当地一个人们钟爱的运动休闲空间，同时为以后潜在的使用功能提供可能性。

Park Landscape
公园景观

KEYWORDS 关键词

Natural & Rustic
自然质朴

Beautiful Environment
环境优美

Peaceful & Quiet
平和宁静

New Chinese Style
新中式风格

Location: Changsha, Hunan
Landscape Design: L&A Design Group
Land Area: 230,000 m²

项目地点：湖南省长沙市
景观设计：奥雅设计集团
占地面积：230 000 m²

Changsha Meixi Lake Peach Blossom Mountain Park
长沙梅溪湖桃花岭山体公园

FEATURES 项目亮点

The overall design of this park manifests the natural life style and the space switching from busy city to quiet and diverse mountain peach orchard. It hopes to purify the mind in rustic nature and advocate healthy values.

公园的整体设计，彰显自然的生活方式与空间切换——由繁忙都市到静谧多样的山地桃园，希望通过自然质朴、净涤身心，提倡健康的价值观。

Overview

Meixi Lake area is an important area in Changsha's new city construction, in which the peach blossom landscape area is with beautiful environment and distinctive characteristics. The peach blossom mountain park, with a land area of 230,000 m², is in the northwest of peach blossom landscape area.

项目概况

梅溪湖片区是长沙新城市建设的重点区域。片区中的桃花岭景区环境优美、特色鲜明。桃花岭公园则位于桃花岭景区西北部，占地达 230 000 m²。

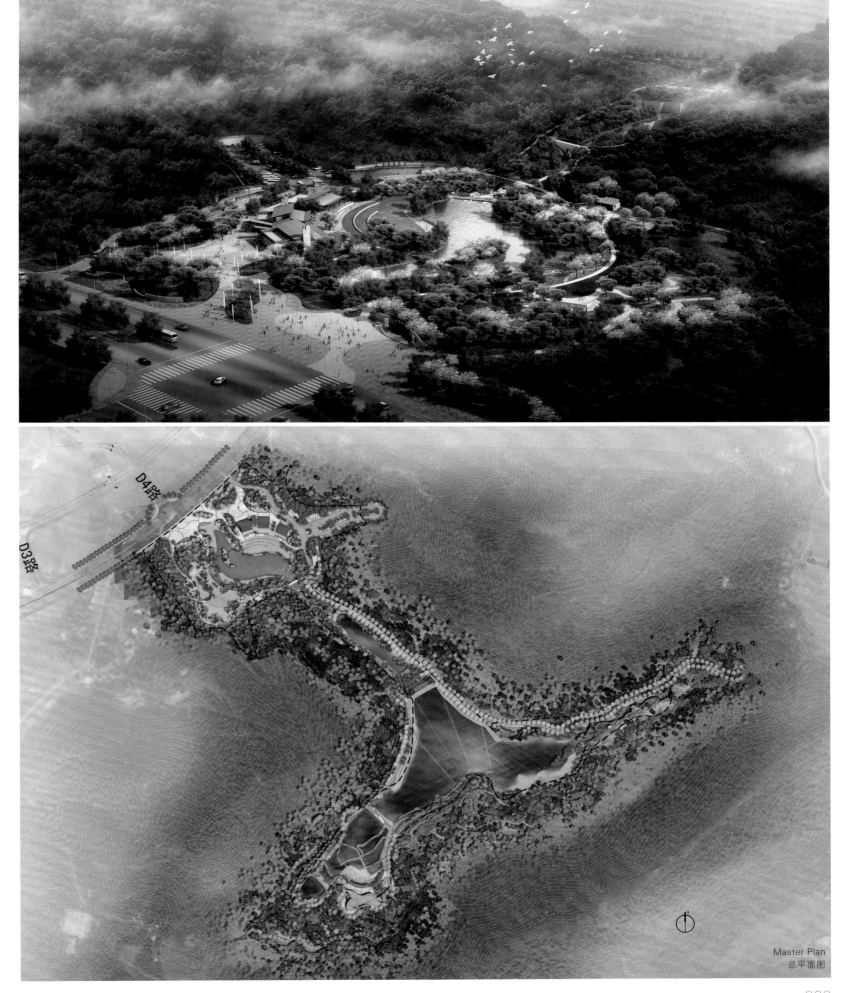

Master Plan
总平面图

Planning Concept

The landscape design remains the mountain style and landscape features of peach blossom landscape area with the vision of "peach garden beside the Meixi Lake, wondering in the peach blossom mountain", and it strives to design a peaceful, quiet, natural, leisurely and open landscape park and embody the reciprocal relationship between human and nature.

规划理念

景观设计延续了桃花岭风景区的山体风貌及景观特征,其设计愿景为"梅溪湖畔桃花源,桃花岭间水云天",力求创造一个平和宁静、自然悠远的开放式山水公园,体现人与自然的互惠关系。

Landscape Design

The park is divided into five functional areas: "the pink peach blossom reception" at the entrance, "forest and stream exploration" in the falling stream landscape area, "clear water and blue sky" in the reservoir landscape area, "colorful peach garden" in the featured waterscape area, and "fragrance in stroll" in the ecological meadow area. The design elements borrow from the local flower shape, drainage form and mountain texture, emphasizing contribution of the peach blossom mountain park to vision of the growing new city and the promotion of life quality. The design obeys the ecological principle of infusing natural ecological landscape into the touring process. It builds a reciprocal relationship between human and nature, and begins with coordinating the harmonious development of ecological, social, cultural, and economic relations from an overall situation, showing respect for historical culture resources and the public.

景观设计

公园分五大功能区：入口区"粉桃迎客"、景观跌溪区"林涧探踪"、景观水库区"镜水云天"、特色水景区"水彩桃源"、生态草甸区"觅香闲步"。设计风格和元素援引地方花型、水系形态和山地肌理，强调桃花岭公园对创造成长中的新城市远景和提升生活品质所做的贡献。设计遵循了将自然生态景观融入游览过程的生态性原则，建立了人与自然的互惠关系，并从协调生态、社会、人文、经济效益关系和谐发展的大局出发，体现了对历史文化资源以及对公众性的尊重。

Park Landscape
公园景观

KEYWORDS 关键词

Native Plants
乡土植物

Eco-friendly
生态环保

8 Areas & 2 Islands
八区两岛

Modernist Style
现代主义风格

Location: Zhangzhou, Fujian
Landscape Design: L & A Design Group
Green Area: 334,639 m²

项目地点：福建省漳州市
景观设计：奥雅设计集团
绿地面积：334 639 m²

Zhangzhou Bihu Ecological Park
漳州碧湖市民生态公园

FEATURES 项目亮点

The planting design highlights the local unique vegetation culture. It uses lots of native plants and creates sustainable vegetation ecosystem, forming a stable group structure.

种植设计凸显了具有当地特色的植被文化，大量使用乡土植物，创建可持续的植被生态系统，形成稳定的群落结构。

Overview

Zhangzhou, located in the "Golden Triangle" zone of Xiamen–Zhangzhou–Quanzhou, has a distinct advantage in location. Bihu Ecological Park is defined as the first-class municipal park in the country and a public waterscape park which shows the ecological landscape planning idea and public leisure features.

项目概况

漳州处于厦漳泉"金三角"地带，区位优势明显。碧湖市民生态公园定义为国内市级公园中一流的、体现生态景观规划理念与公共休闲特色的市民水景公园。

01	百花园
02	雕塑园
03	浦头港入口
04	阳光草坪、滨湖步道
05	碧湖港
06	都市入口
07	碧湖湾、渔人码头
08	儿童乐园
09	康乐园
10	公园展示中心
11	过街、过江天桥
12	江滨路入口
13	景观湿地
14	自然水岸、江滨路
15	儿童科普天地
16	水仙岛
17	台地花园
18	文化水岸
19	生态岛
20	滨江运动公园

Site Plan
总平面图

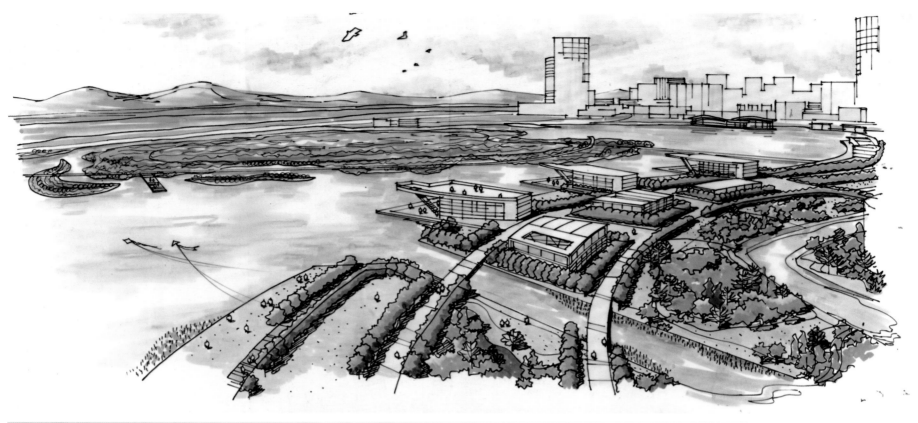

| 密林 | 儿童活动区域 | 道路 | 绿林缓坡 | 水道 | 生态湿地 | 水道 | 生态湿地 | 水道 | 生态湿地 | 水道 | 生态湿地 | 观景平台 |

| 广场 | 休息平台 | 广场 | 滨水漫步道 | 驳船码头 | 木栈道 | 滨水漫步道 | 商业建筑 | 滨水活动广场 |

Development Idea

The design highlights the ecological, natural, low-carbon and environmental idea, respects the local culture features, motivates business activities and advocates community culture. It builds wetland ecosystem, making water be the organic connection and belt of the whole park system and various functions and become the soul of the park. It uses ecological economy and tourism economy concept as guidance, obeys the principle of combining development and protection and combines various functions to offer people a place for travel, vacation, leisure, education and cultural entertainment, making the ecological park become a green and ideal place for the citizens.

开发思路

项目突出生态自然与低碳环保的理念，尊重地方文化特色，激发商业活力，弘扬社区文化；建立湿地生态系统，使水体成为公园整个系统及各种功能实现的有机联系和纽带，成为公园的灵魂；以生态经济和旅游经济理论为指导，遵循开发与保护相结合的原则，将多种功能相结合，为人们提供旅游、度假、休闲、教育、文化娱乐的场所，使生态公园成为市民的绿色梦想之地。

Planning Concept

The design uses traditional Chinese silk ribbons "showing the beauty of life with dancing ribbons" as conceptual theme, showing how it strengthens the relationship between life and culture, city and nature, and continuing the stories of the past and the future. Three ribbons surround the park and respectively represent natural landscape, urban landscape and cultural landscape, forming the content of this project. Each ribbon connects and manifests the curved shoreline, corresponding functional facilities and activity contents.

规划理念

以传统的中国丝绸缎带"生活之秀舞飘带"为概念主题，演绎如何加强生活与文化、城市与自然之间的关系，承接过去和未来的故事。三抹缎带环绕公园，分别代表自然景观、城市景观和文化景观，建构项目内容。每幅条带关联并凸显曲水岸线及相应的功能设施和活动内容。

Landscape Planning

According to the planning layout, the ecological park can be functionally divided into 8 areas and 2 islands that are correlative and distinctive. The 8 areas are fashion leisure area, lakeside business and sunshine lawn area, healthy and happy activity area, park exhibition area, wetland landscape area, fishing leisure and science education area, Jiangbin Road landscape area and lakeside cultural corridor; the 2 islands are ecological island and narcissus island. Each area combines with the adjacent relationship of different urban functional areas, fully integrating into the civil life and permeating the natural ambiance into the butted urban space.

景观规划

根据规划布局，生态公园可划分为"八区两岛"，相互关联各具特色的功能区域。八区：时尚休闲区、滨湖商业与阳光草坪区、康乐活动区、公园展示区、湿地景观区、垂钓休闲与科普教育区、江滨路景观带、湖滨文化长廊。两岛：生态岛、水仙岛。各区与城市不同功能区块邻接，充分融入市民生活，并使自然气息渗透到对接的城市空间中。

Ecological Design

The biggest problem of Bihu Ecological Park is water treatment. To use ecological, energy saving and economical methods to solve the landscape water quality, water yield, flood control and drainage problems, as well as to improve landscape image and municipal functions of sewage treatment plant is always the main concern of this design. The designer leads the water of Chiu-lung River into the Bihu ecological wetland island group for purification and filtration and combines ecological grass ditch design to collect and gather treated rainwater into the lakes for landscape water supplement. Besides, part of the lakes can use the upgraded water from the sewage treatment plant. The water of Putou Harbor would be drained into Chiu-lung River through the lake's bottom elevation design and pipe after treatment.

The planting design highlights the local unique vegetation culture. It uses lots of native plants and creates sustainable vegetation ecosystem, forming a stable group structure. The vegetation group structure shows the development from suburb to urban and the evolution from wetland ecosystem to terrestrial ecosystem, which is ecological, ornamental and scientific.

The overall design applies lots of ecological engineering and facilities, such as constructed wetland, wetland floating island, rainwater garden, ecological open trench, green roof, wall and so on. Through the application of different ecological facilities, this park becomes a model of ecological and cultural combination.

生态设计

碧湖生态园开发中遇到的最大问题是水处理问题。如何用生态、节能和节约的方法，解决景观水质、水量和防洪排涝问题，以及改善污水厂景观形象和提升其市政功能，始终是设计关注的要点。设计师引九龙江水进入碧湖生态湿地岛屿群中进行净化和过滤，并结合生态草沟设计，引导收集和汇集处理雨水至湖体，补充景观用水。另外，部分湖体的水源也可来源于污水处理厂的升级改造。铺头港的水通过湖底高程设计和管道，处理后排入九龙江。

种植设计突显了具有当地特色的植被文化，大量使用乡土植物，创建可持续的植被生态系统，形成稳定的群落结构。植物群落结构展示了郊野到城市的发展，湿地生态系统到陆地生态系统的演化，具有生态性、观赏性和科普性。

整体设计运用诸多的生态工程和生态设施，包括人工湿地、湿地浮岛、雨水花园、生态明沟、绿色屋顶与墙面等。通过不同生态设施的使用，使公园成为生态与人文结合的典范。

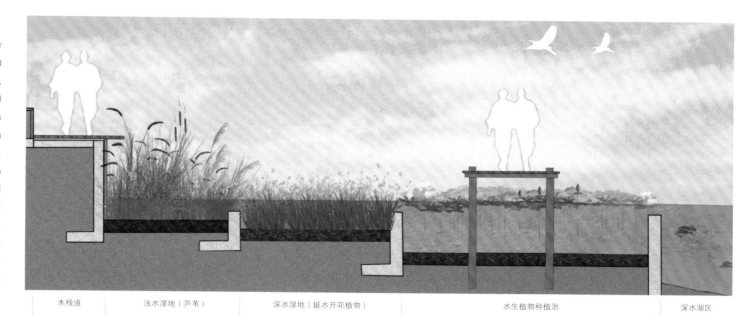

木栈道　　浅水湿地（芦苇）　　深水湿地（挺水开花植物）　　水生植物种植池　　深水湖区

| 绿化 | 儿童娱乐设施场地 | 休闲平台 | 亲水石滩 |

| 绿化 | 休闲步道 | 特色景墙 | 儿童娱乐设施场地 | 休闲平台 | 亲水石滩 |

Commercial Landscape

商业景观

Urban Ecology
城市生态

Landscape Elements
景观小品

Leisurely Space
休闲空间

Commercial Landscape
商业景观

KEYWORDS 关键词

Theme Landscape
主题景观

Landscape Node
景观节点

Green Ecology
绿色生态

Modernism Style
现代主义风格

Location: Chaoyang District, Beijing
Developer: SOHO China Limited
Landscape Design: Ecoland
Land Area: 115,393 m²
Floor Area: 521,265 m²
Green Coverage Rate: 30%
Plot Ratio: 8.14
Year of Completion: End of 2014

项目地点：北京市朝阳区
开发单位：SOHO 中国有限公司
景观设计：ECOLAND 易兰规划设计院
占地面积：115 393 m²
建筑面积：521 265 m²
绿化率：30%
容积率：8.14
建成时间：2014 年底

Wangjing SOHO
望京 SOHO

FEATURES 项目亮点

Following the architectural design idea, the designers fully integrate functions and form in the landscape planning and design, making the 50,000 m² landscapes dialogue with the buildings full of visual and spatial dynamic effect, and allowing them to complement each other.

沿用扎哈在建筑设计中运用的独特曲面造型，易兰设计师在项目景观规划，使项目 5 万 m² 的超大景观园林，与极具视觉和空间动感效果的建筑形成对话，相得益彰。

Overview

The project is located in the second CBD of future Beijing — Wangjing core area with broad vision, overlooking Wangjing and extending east to Futong West Street, south to Fu'an East Road, west to Wangjing Street and north to Fu'an West Road.

项目概况

项目位于未来北京的第二个 CBD——望京核心区，视野开阔，俯瞰望京。东至阜通西大街、南至阜安东路、西至望京街、北至阜安西路。

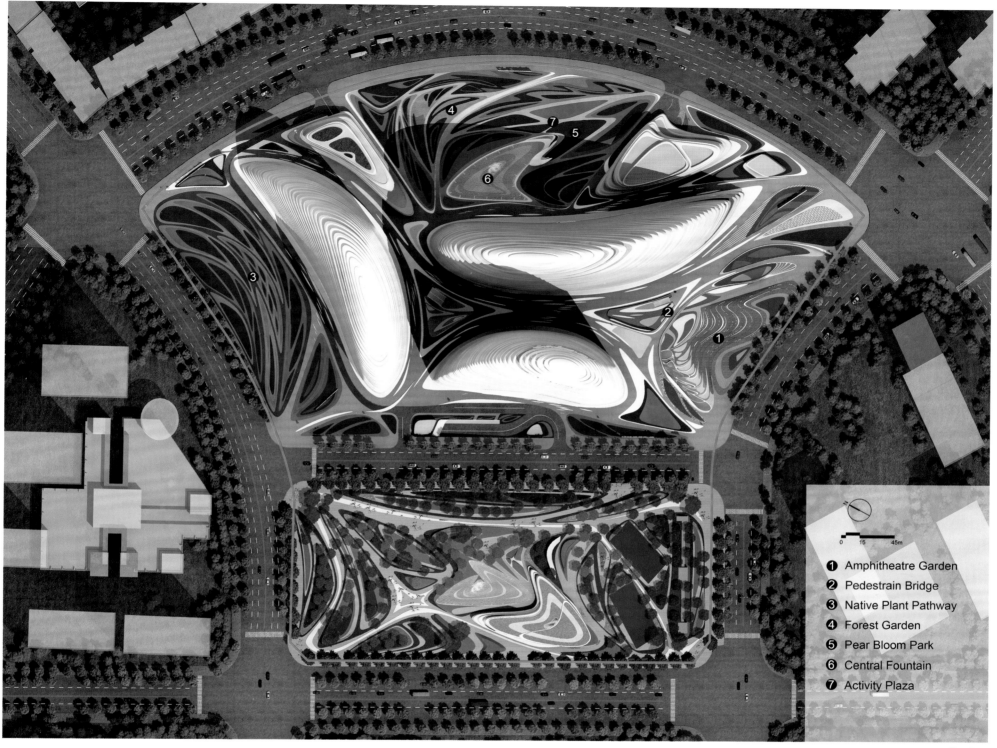

Site Plan
总平面图

Positioning Strategy

Initial project planning is a sub-center of city close to Olympic Rings and dominated by residences which can accommodate 600,000 people. With urban development, Wangjing gradually becomes a region gathering IT companies. The establishment and construction of the Big Wangjing in the north make the project suddenly become Beijing's urban business district nearest to the airport.

Located on the central axis of the project plan, plot B29 parallels to the airport expressway, back against the Big Wangjing and facing to Wangjing Central Park in the south side. Thus this particular center position and relationship with the surrounding buildings decides that plot B29 is going to become the core and landmark of Wangjing.

定位策略

项目最初规划是一个紧靠五环,以居住为主的能容纳60万人口的城市副中心。随着城市发展,望京逐渐成为IT公司聚集的区域。北侧大望京的确立和兴建,使项目顿然成为距离机场最近的北京城市商务区。

B29地块处在项目规划的中轴线上,与机场高速平行。背靠大望京,南侧面对望京中央公园,这一特殊的中心位置和与周边楼宇的关系,决定了B29地块必然成为望京的核心和地标。

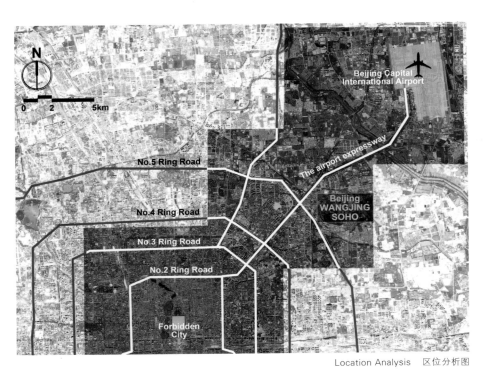

Location Analysis　区位分析图

Landscape Structure　景观结构图

Planning and Layout

Project consists of three high-rise buildings integrating office and commerce and three single-family low-rise commercial buildings, among which one reaches the maximum height of 200 m. Since its completion in 2014, Wangjing SOHO becomes "the first impression building of the capital", and the first attractive high-rise landmark after entering downtown from the Capital Airport.

Design Concept

The domestic top design institution Ecoland and Zaha Hadid Architects cooperate to perfectly combine and fully display their design style and strength from architectural design to landscape design. Unique surface modeling makes the buildings at any point of view present dynamic and elegant beauty. Tower exterior is covered with flashing aluminum plate and glasses, blending with the blue sky and symbolizing the vision of misty mountains.

规划布局

项目由3栋集办公和商业一体的高层建筑和三栋低层独栋商业楼组成，最高一栋高度达200 m。2014年建成后望京SOHO成为了"首都第一印象建筑"，是从首都机场进入市区的第一个引人注目的高层地标建筑。

设计理念

由国内顶级设计机构ECOLAND易兰规划设计院与扎哈·哈迪德（Zaha Hadid）建筑事务所倾力合作，从建筑设计到景观设计，双方设计风格和实力得到了完美的结合和充分展现。项目建筑独特的曲面造型使建筑物在任何角度都呈现出动态、优雅的美感。塔楼外部被闪烁的铝板和玻璃覆盖，与蓝天融为一体，象征了山中云雾缭绕的意境。

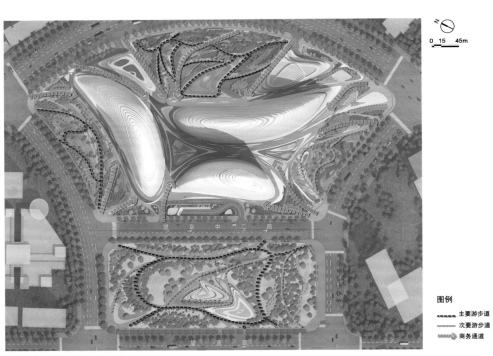

Traffic Drawing　交通分析图

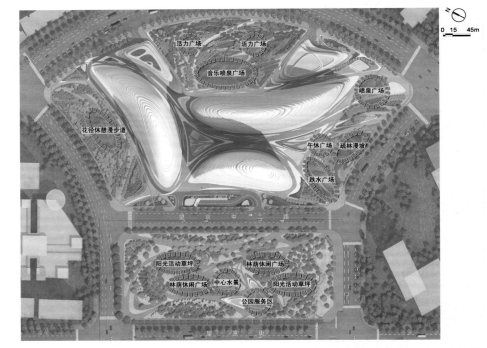

Functional Areas Diagram　功能分区图

Landscape Design

Following the architectural design idea, the designers fully integrate functions and form in the landscape planning and design, making the 50,000 m² landscapes dialogue with the buildings full of visual and spatial dynamic effect, and allowing them to complement each other. Green coverage rate of 30% forms a unique urban garden-style office environment. In order to reflect the change of seasons, Ecoland design team offers Wangjing SOHO with four theme landscapes as leisure theatre, sports ground, art sculptures and waterscape. The distinctive musical fountain and garden landscape look like the sunny garden when seen from afar and complement with the buildings. All these make the whole project in terms of architecture, landscape and construction organizations have reached the US Green Building LEED certification standard, and created an energy-efficient, water-saving, comfortable and intelligent green building in Beijing.

景观设计

沿用扎哈在建筑设计中运用的独特曲面造型,易兰在项目景观规划设计中将功能与形式充分结合,使项目5万m²的超大景观园林,与极具视觉和空间动感效果的建筑形成对话,相得益彰。绿化率高达30%,形成了独树一帜的都市园林式办公环境。为了体现四季更迭变化,易兰设计团队为望京SOHO特别打造了休闲剧场、场地运动、艺术雕塑、水景四大主题景观。其独具匠心的音乐喷泉和园林景观,远远望去如同洒满阳光的花园,与楼群相辅相成。这一切使得整个项目在建筑、景观和施工组织等方面都达到美国绿色建筑LEED认证标准,打造出一个节能、节水、舒适、智能的北京新绿色建筑。

Commercial Landscape
商业景观

KEYWORDS 关键词

Lawn space
草坪空间

Landscape Trails
景观步道

Theme Plaza
主题广场

Modernism Style
现代主义风格

Location: Nanchang, Jiangxi
Developer: Greenland Group
Landscape Design: W&R GROUP
Size: 17,250 m²

项目地点：江西省南昌市
开发商：绿地集团
景观设计：水石国际
项目规模：17 250 m²

Greenland International Expo City, Nanchang
南昌绿地博览城

FEATURES 项目亮点

The main idea of this Expo City's residential project is product series with the theme of "Jade", so the design should display humanistic, artistic, healthy, technological and ecological residential culture based on the overall idea.

整个博览城住宅项目主打理念为"玉"主题产品系，设计需在整体理念背景下，展现人文、艺术、健康、科技、生态的居住主题文化。

Overview

This case is the sales offices of Greenland Nanchang Expo City Industrial New Town residential project. How to express the different styles of the Expo City's residential properties and display the interpretation and high quality pursuit of Nanchang Greenland in terms of residential culture is the focus of this design.

项目概况

本案为绿地南昌博览城产业新城住宅项目的售楼处，如何表达博览城不同住宅产品系的风貌，及绿地南昌对于居住文化的解读和高品质追求是本案设计的重点。

	特色 logo
	特色浅水镜面水池
	特色点景挡墙
	营销活动广场
	生态停车场地
	阳光观赏草坪
雅宇	
样板区入口	逸宇
竹子围合边界	景墙结合路径
逸宇	玺宇
洋房内庭院	中心庭院景观
	睿宇

Site Plan
总平面图

Design Concept

The main idea of this Expo City's residential project is product series with the theme of "Jade", so the design should display humanistic, artistic, healthy, technological and ecological residential culture based on the overall idea. The designers begin with the overall idea of healthy living to create sales offices landscape that can accommodate both complex functional requirements and specific cultural theme.

设计理念

整个博览城住宅项目主打理念为"玉"主题产品系,设计需在整体理念背景下,展现人文、艺术、健康、科技、生态的居住主题文化。以健康生活的整体概念出发打造既容纳复合功能需求,又涵盖特定文化主题的售楼处景观。

Landscape Layout

The overall layout consists of sales office landscape space and showflats' garden space.

Sales Office Landscape Space

The front plaza entrance combines architectural facade in landscape theme to set up the mirror waterscape, which reflects the environment relationship between the building and the surroundings to form multiple landscape interfaces, creating the marketing exhibition space full of rich cultural theme in the front square and the display interface along the road.

Featured logo integrated with landscape wall guides people flow through the ribbon trail in layers to arrive the sales office entrance space. Featured stainless steel folding landscape wall combined with dry spray, thematic show light box to form a rich landscape display wall and constitute the sales offices entrance theme square together with lawn space, characteristic landscape trails and displaying square.

Model House Displaying Garden Space

Building the theme concept of the wall and the house.

Wall: concept of hectometer-long scroll showing the charm of Jiangxi's building is continued to define interior and exterior space with the concept of wall. The guide and obstruction of the inner ring of walls form a continuous circular path, while the outer ring of walls combine with inner wall to form multiple spaces.

House: emphasizing the wall enclosure and permeability to form unlimited extension of the space and the central garden and the surrounding five theme culture display courtyard: Xi Yu (culture), Yi Yu (leisure), Ya Yu (art), Lan Yu (nature), Rui Yu (science and technology).

The concept of combining the central courtyard with the Japanese rock garden forms the ornamental-oriented cultural theme display area. The surrounding five display courtyards combine with the outdoor negotiation, children's entertainment display, walking and exercising path display and other leisure function theme, forming the rich experience of the interesting space. Clients constantly feel and deepen the living cultural identity and understanding of the sales offices in the space where sceneries change by walking, conceiving life scenes of the future to produce a sense of belonging.

景观布局

整体布局上由售楼处景观空间和样板房展示庭院空间两部分组成。

售楼处景观空间

入口前广场结合建筑的山水主题立面设置镜面水景，水镜反射建筑及周边的环境关系，形成多重山水界面，营造出文化主题浓郁的前广场营销展示空间及道路展示界面。

与景墙一体化的特色 logo 引导人流通过带状的层级步道到达售楼处入口空间，特色的折面不锈钢景墙结合旱喷、主题展示灯箱形成丰富的景观端景展示墙。与草坪空间、特色景观步道、展示广场一起构成售楼处入口主题广场。

样板房展示庭院

打造墙宇的主题概念。

墙：延续建筑展现江西魅力百米长卷的概念，用墙的概念界定内部、外部空间，内圈墙体的引导和阻隔形成连续的环形路径，外环的墙体结合内墙形成多重空间。

宇：强调墙体的围合与渗透形成无限延伸的空间，形成中心的墙宇庭院及周边五个主题文化展示庭院：玺宇（人文）、逸宇（休闲）、雅宇（艺术）、澜宇（自然）、睿宇（科技）。

中心庭院结合枯山水的概念形成了观赏为主的文化主题展示区，周边五个展示庭院结合户外洽谈、儿童娱乐展示、漫步健身路径展示等停留休闲功能主题，形成体验丰富的趣味空间，客户在步移景异的空间中不断感受、加深售楼处对于居住文化的认同和理解，构想未来的生活场景，从而产生归属感。

Commercial Landscape
商业景观

KEYWORDS 关键词

Green Space
绿化空间

Diversified
变化多样

Podium Landscape
裙楼景观

Modernism Style
现代主义风格

Location: Guangzhou, Guangdong
Client: Guangzhou Zhanhui Real Estate Development Co., Ltd.
Landscape Design: Keymaster Consultant Co., Ltd.
Design Team: Rowan, Sarah Choi, Gao Lijuan, Lin Jianxin
Land Area: 10,000 m²

项目地点：广东省广州市
业主：广州展汇房地产开发有限公司
景观设计：广州市科美都市景观规划有限公司
设计团队：罗文、蔡舒雁、高丽娟、林健欣
占地面积：10 000 m²

Landscape Design for La Vendome, Pazhou, Guangzhou
广州市琶洲南丰汇景观设计

FEATURES 项目亮点

Linear green spaces are staggered and interwoven to form a mult-level landscape system together with the mirror-like water features. And the skillful lighting design has further enhanced the commercial atmosphere.

线性的绿化空间相互重叠、交错，结合镜面水景，使绿化空间收放有致、层次变化多样，在灯光的衬托下，更添商业气氛。

Overview

Located on Xingang East Road of Haizhu district, to the south of Pazhou International Convention & Exhibition Center, the project is built above Metro Line 2, being one of the landmarks in Pazhou.

项目概况

项目位于海珠区新港东路，琶洲会展中心南侧。地铁二号线上盖物业，是琶洲的地标性项目。

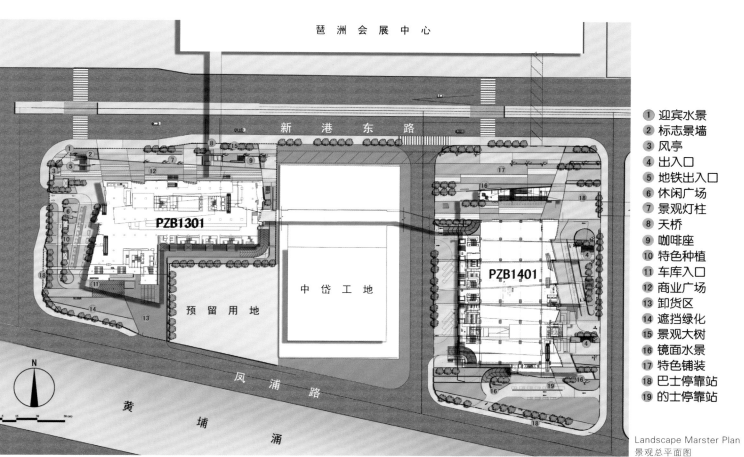

① 迎宾水景
② 标志景墙
③ 风亭
④ 出入口
⑤ 地铁出入口
⑥ 休闲广场
⑦ 景观灯柱
⑧ 天桥
⑨ 咖啡座
⑩ 特色种植
⑪ 车库入口
⑫ 商业广场
⑬ 卸货区
⑭ 遮挡绿化
⑮ 景观大树
⑯ 镜面水景
⑰ 特色铺装
⑱ 巴士停靠站
⑲ 的士停靠站

Landscape Marster Plan
景观总平面图

Design Idea

Quadrangle composed of four straight lines is the basic landscape element. Quadrangles, parallelograms, triangles and polygons of different sizes are well connected and cleverly organized in the framework to create different functional spaces. Important nodes are also strengthened by those lines. It is a new composition that's full of changes. Linear green spaces are staggered and interwoven, together with the mirror-like water features to form a mult-level landscape system. And the skillful lighting design has enhanced the commercial atmosphere.

The landscape of the podium also applies geometric framework. Linear green spaces cover the roof to form natural forest roofing which keeps harmonious with the ground floor square when seen from high.

The stacked buildings and the linear green spaces are combined and interplay with each other to create an integrated whole.

设计理念

以直线组成的四边几何形为构成的基本形,再重复骨骼中输入大小不等的四边形、平行四边形、多角形,但保持骨骼内各单位间一定的空间关系,体块显示出最后的功能分区,空间划分。同时重点节点部分也在线条运用中得到了加强。这也是一种富有变化的新的构成方式。

线性的绿化空间相互重叠、交错,结合镜面水景,使绿化空间收放有致、层次变化多样,在灯光的衬托下,更添商业气氛。

裙楼景观同样延续几何形的框架,线性的绿化覆盖形成了天然的森林屋面,这种形态从高层看下来,与首层广场协调统一。

设计将建筑的重叠与绿化的线条相互影响,和谐互动,形成统一的整体。

Commercial Landscape
商业景观

KEYWORDS 关键词

Lighting Effect
灯光效果

Commercial Atmosphere
商业氛围

Theme Sculpture
主题雕塑

Modernism Style
现代主义风格

Location: Changshu, Jiangsu
Client: Changshu World Fashion City
Landscape Design: Dongda Landscape Design
Size: 27,000 m²

项目地点：江苏省常熟市
委托单位：常熟商业广场服装城
景观设计：深圳市东大景观设计有限公司
项目规模：27 000 m²

Changshu Merchants Culture Commercial Plaza
常熟招商文化商业广场

FEATURES 项目亮点

The design adopts the landscape layout of "one ring, two axes, and three points" to build a modern, high quality and local plaza integrating shopping, tourism and leisure.

设计采用"一环、两轴、三点"的景观布局结构，旨在打造一个现代的、高品质的、具有地方特质的综合性购物旅游休闲广场。

Overview

The project, located in Changshu Merchants City of southern Changshu City which formerly is the Changshu Merchant City Xinfeng small commodity market, is the centeral region of Merchant City with convenient traffic, crowded people flow and strong commercial atmosphere.

项目概况

项目位于常熟市南部常熟招商城内，原为常熟招商城新丰小商品市场，是招商城的中心区域，交通便捷，人流密集，商业气氛浓郁。

Landscape Layout

The design adopts the landscape layout of "one ring, two axes, and three points" to build a modern, high quality and local plaza integrating shopping, tourism and leisure. One ring, with center stage in the center, is formed through the sinking square, strengthens the contact between square and the underground space, and enables visitors in the square quickly get into the underground commercial space and parking lot. Two axes, namely Star Boulevard and landscape secondary axis, become an important basis for landscape connection, sight line and traffic streamline organization. Three points: the entrance theme sculpture, characteristic backdrop and center stage mutually match. Large folding steel backdrop by its unique modelling highlights the fashion sense and sense of science and technology and greatly strengthens the uniqueness and the landmark of the project.

Center Stage Area

Center stage is located in the center square and connected with backdrop in the same main axis. Here can hold a grand performance, in the evening, the stage and the lights in backdrop add radiance and beauty to each other, so that the center stage will be the most dazzling place in the plaza.

Without affecting the use of region space function, center stage area will be designed as dual-use places for show and conference,

景观布局

设计采用"一环、两轴、三点"的景观布局结构,旨在打造一个现代的、高品质的、具有地方特质的综合性购物旅游休闲广场。一环:以中心舞台为中心、通过下沉广场形成,加强了广场和地下空间的联系,使游人在广场的各个方位都能快捷地进入地下商业空间及停车场。两轴:星光大道及景观次轴,成为景观串联与视线、交通流线组织的重要依据。三点:入口主题雕塑、特色天幕和中心舞台,相互烘托,大型的折形钢构天幕以其独特的造型凸显时尚感与科技感,大大强化了该项目的独特性和地标性。

中心舞台区

中心舞台位于广场的中心地带,与天幕位于同一个主轴上。由星光大道将两者相连。在这里可以举行盛大的演出。在晚上,舞台和天幕的灯光交相辉映,中心舞台将会成为广场最耀眼的地方。

在不影响空间区域功能利用的情况下,设计将中心舞台区设置为表演兼会议

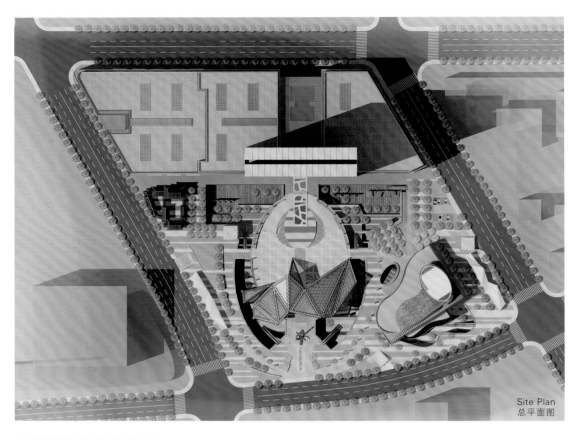

Site Plan
总平面图

fully making use of effective space of the square to meet the functional requirements.

During the performance, the actors get into the center stage through T-shaped stage. Watching district and platform form a semi-open theatre space. When holding a meeting, adequate hard ground and structured venue would be an ideal choice. Usually it is a large venue that provides possible activities for people and fully meets the use of leisure and entertainment activities. Some mobile tree pool is placed in usual time to increase the green area and beautify the square landscape.

Building in the triangle basic form creates large scale backdrop by overlaying and combination. On the one hand, it exists as a primary symbol of the Cultural Square to emphasize the importance of the function and symbol, on the other hand, it plays a shielding role in function, allowing visitors to stay and rest for a long time.

的两用场所，充分利用广场的有效空间，也满足了功能上的需求。

在进行演出时，演员经过舞台T台进入中心舞台，观演区及观台形成一个半开敞的观演空间。举行会议时，场地充足的硬质地面及层次分明的场地不失为一个理想之选。平时它又是一个大型的活动场地，为人们提供了充分的活动可能，充分满足人们休闲、娱乐活动之用，平时摆放一些移动式树池，以增大广场的绿化面积，美化广场的景观。

建筑以三角形为基本形，通过叠加、组合，形成大体量的天幕，它一方面做为文化广场的首要标志物存在，强调出广场功能的重要性、标志性，另一方面它在功能上起到了遮挡的作用，让游人得以长时间的停留、休憩。并通过夜景灯光的处理，使其在晚间形成多样的夜景灯光效果。

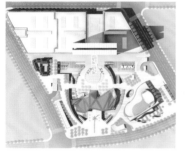

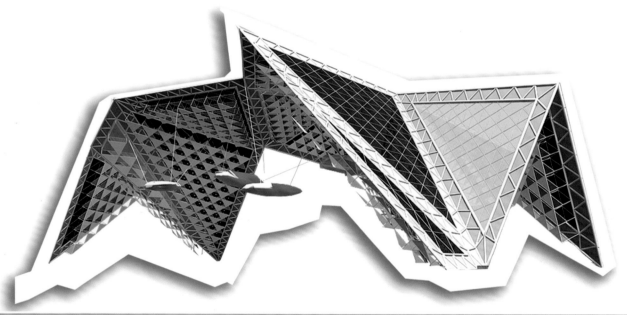

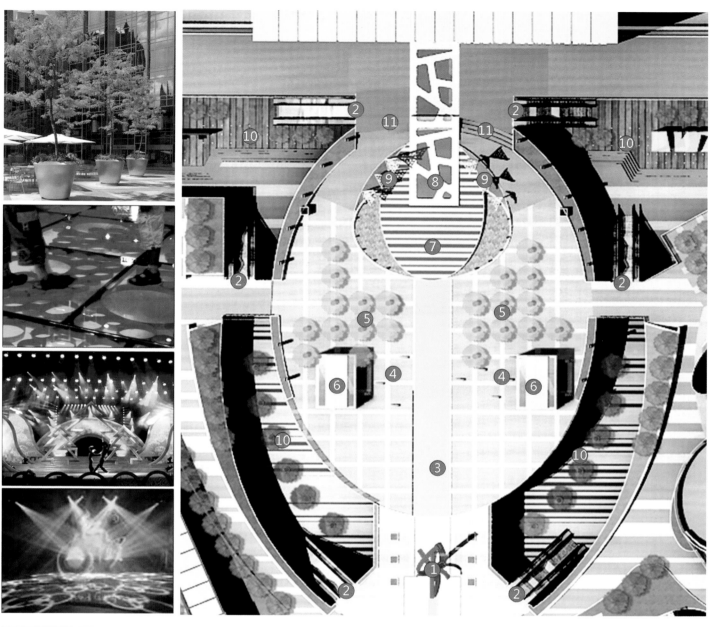

1. 景观雕塑
2. 自动扶梯
3. 星光大道
4. 特色座椅
5. 移动树池
6. 人防出入口
7. 中心舞台
8. 舞台T台
9. 特色舞台构架
10. 下沉广场
11. 台阶

中心舞台平面图

Commercial Landscape
商业景观

KEYWORDS 关键词

Landscape Wall
景墙设计

Theme Sculpture
主题雕塑

Landscape Node
景观节点

Modernism Style
现代主义风格

Location: Huadu District, Guangzhou, Guangdong
Client: Guangzhou K Wah Huadu Property Limited
Landscape Design: Keymaster Consultant (Guangzhou) Ltd.
Designer: Zhang Zhaoyun, Liang Guozhi, Huang Wanjun, Chen Jin'en
Area: Appr. 30,000 m²

项目地点：广东省广州市花都区
委托方：广州市嘉华花都置业有限公司
景观设计：广州市科美都市景观规划有限公司
设计师：张兆云、梁国智、黄婉君、陈晋恩
面积：约 30 000 m²

K Wah Crowne Plaza
嘉华皇冠假日酒店

FEATURES 项目亮点

The design concept of "crystal seasons" in the project uses crystal sculpture and stainless steel to reflect ambient light which forms four seasons' hue changes and to make it become the embodiment of the hotel's unique personality and noble quality.

以"水晶四季"作为项目的设计概念，利用晶体雕塑及不锈钢材质折射环境光形成四季色相变幻，使其成为酒店的独特个性与尊贵品质的体现。

Overview

The Project is located in Huadu New District, at the junction of Yingbin Avenue and the Qingpu Avenue. Two phases of the development consist of phase I for the hotel and office land with an area of about 30,000 m², and phase II for residential.

The investor of the hotel is Hong Kong K Wah Company, and the hotel was taken over by IHG Hotel Management Company when built and entitled Crowne Plaza Hotel. The project's orientation is the modern hotel landscape style in harmony with the architectural style, and is committed to building a local landmark five-star business hotel.

项目概况

项目地处花都新区，位于迎宾大道与清埔大道的交汇处。分两期开发，一期为酒店及写字楼用地，面积约 30 000 m²，二期为居住小区用地。

酒店的投资方为香港嘉华公司，建成后由 IHG 酒店管理公司接管并题名为皇冠假日酒店。项目定位是与建筑风格相协调的现代酒店式园林风格，并致力打造成当地地标的五星级商务酒店。

Design Concept

The design concept of "crystal seasons" in the project uses crystal sculpture and stainless steel to reflect ambient light forming four seasons' hue changes and makes it become the embodiment of hotel's unique personality and noble quality. At the same time, landscaping accessory lines created by the use of human visual, auditory, olfactory and tactile sense enrich different sensory spaces on the way.

设计概念

以"水晶四季"作为项目的设计概念,利用晶体雕塑及不锈钢材质折射环境光形成四季色相变幻,使其成为酒店的独特个性与尊贵品质的体现。同时通过利用人的视觉、听觉、嗅觉及触觉作为造景的辅线,丰富不同的沿路感官体现空间。

The Main Landscape Nodes

Main Entrance Area

The main entrance landscape area is between the sidewalk on the north side of Yingbin Avenue and the main hotel building. But the main problem under the present situation is to avoid a 6 m wide municipal canal extending from east to west in front of the hotel; meanwhile to meet the requirements of not moving trees within 15 m wide municipal green belt made by the government. These two issues will directly block the hotel facade by messy green and limit landscape shaping by irrigation ditch. The owners communicate with the government for many times and fail to properly solve the problem. Finally, the designer has a comprehensive consideration to the proposal. Two simple waterscape walls combine to be the front landscape of the hotel. The landscape wall shapes as if a group of scenery fell down from the building. Facade design continues the vertical design elements of external appearance. Pavement, landscape wall and building are integrated as a whole.

Back Garden Area — Theme Sculpture Platform

Theme sculpture platform in the back garden area as the main landscape node against the lobby lounge is the top priority of the whole area. Here is the focus point of the hotel landscape visual corridor, so its scale, distance, and shape are required to make repeated scrutiny. In addition, the materials of sculpture and the design of the pedestal also need to consider.

Designers use superposition of glass to shape into irregular crystals, whose outer surface is polished. The pedestal floating on the water surface and the sculpture combine to make the ice-crack-shaped designs and the crevices between stones also are embedded by stainless steel material to echo with the main entrance. The lighting is designed through the underwater light refraction to project into the inner wall of the sculptures, making it a lighting group with changing landscape.

Back Garden — Swimming Pool

Earlier, the pool is to be considered as an indoor swimming pool, and later due to changes in the requirements of the owner, it was converted into an outdoor pool. But the structure and shape of the landscape has been completed, the re-intervention cannot make a big adjustment, therefore, in the design, the designer considers more of veneer, details and the surrounding landscape.

As an open landscape swimming pool, designers

hope to turn it into a simple and comfortable leisure space, and the bottom of the mosaic pattern is still a continuation of the architectural elements of vertical strip. Combination of strip and square mosaic forms a rich sense of texture. The overflow ditch design hidden surrounding the pool forms the effect of mirror like water. The Terminaliamantaly Triciolor tree rows and small green bamboos become a green background of the pool.

主要景观节点

主入口区

主入口景观区的用地范围为迎宾大道北侧人行道至酒店主体建筑之间。但是就当时现状情况而言存在的主要问题是：酒店门前有一条东西走向6 m宽的市政灌渠需要避让；同时政府方面也提出市政15 m宽的绿化带内的乔木不能移动的要求。这两个问题将直接造成现状杂乱绿化对酒店门面的遮挡及灌渠对景观塑造的限制。经业主与政府的多次沟通，问题并未能得到妥善的解决。最终，设计师对方案进行了综合的考虑。以两面简洁的水景墙交错结合作为酒店的门前景观，景墙造型仿佛从建筑跌下的一个组景，立面的设计延续了建筑外立面竖向的设计元素。铺装、景墙、建筑三者融为一体。

后花园区——主题雕塑台

后花园主题雕塑台作为酒店大堂吧对出的主要景观节点，是整个区域的重中之重。这里是酒店景观视线通廊的聚集点，其尺度、距离、形态均需作反复的推敲。此外，雕塑的选材及底座的设计也有需要斟酌的地方。

设计中设计师以片层叠加的玻璃塑造成不规则的水晶体，晶体的外表面作打磨处理。浮于水面的底座与雕塑相结合作冰裂形设计，石缝之间也嵌入不锈钢材质与主入口相呼应。而灯光的设计是通过水底灯对流水的折射，把光影投射到雕塑的内壁，使其成为一组变幻的灯光组景。

后花园区——泳池

前期泳池是作为室内泳池进行考虑，后来由于业主方的变更要求，将其改为室外泳池。但是结构及形状已经完成，景观的重新介入不能作过大的调整。因此，在设计中，设计师考虑更多的是饰面、细部及周边景观的设计。

作为一个开放式的景观泳池，设计师希望能将其打造成一个简洁而舒适的休闲空间，池底的马赛克图案仍然延续了建筑竖向条带的元素。长条形与方形马赛克的组合，形成一种富有韵律感的肌理，泳池周边隐藏式的溢水沟设计，使其形成镜面水的效果。周边的锦叶榄仁树阵及小青竹成为泳池的一个绿色背景。

Commercial Landscape
商业景观

KEYWORDS 关键词

Drop Water Landscape
跌水景观

Sculpture and Fountain
雕塑喷泉

Terracotta Pantile
红陶筒瓦

Mediterranean Style
地中海风格

Location: Huadu District, Guangzhou, Guangdong
Owner: Guangzhou Mayland Group
Landscape Design: Keymaster Consultant (Guangzhou) Ltd.
Co-designer: Belt Collins
Project Team: Zhu Lei, Feng Zhuowei, Luo Lidan
Landscape Size: 27,000 m²

项目地点：广东省广州市花都区
业主：广州美林基业集团
景观设计：广州市科美都市景观规划有限公司
合作设计：贝尔高林
项目团队：朱蕾、冯焯伟、骆丽丹
景观规模：27 000 m²

Mayland Lake Hot Spring Hotel
美林湖温泉大酒店

FEATURES 项目亮点

Many Mediterranean style hot spring swimming pools are built down along the mountain. All the hotel landscape details will follow the architectural language of the main buildings. Manual plastering by STUCCO, arched doorways and windows, terra cotta pantile, iron art, logs and other extremely delicate techniques create authentic Mediterranean-style resort landscape.

多个依山势而建的地中海风情温泉泳池呈跌级式排列，所有酒店景观的细节都遵循建筑主体的语言，以STUCCO手工抹墙、拱形门廊窗户、红陶筒瓦、铁艺、原木等极为考究的细腻手法，营造出原汁原味地中海式度假酒店景观。

Overview

Mayland Lake Hot Spring Hotel is located in the core area of China Mayland Lake – Hotel theme park, with favorable natural condition of over 30,000 acres of native forests, lakes, springs and golf. The main building is built along the mountain as the theme of "Italy mountain city" to start the project design.

Many Mediterranean style hot spring swimming pools are built down along the mountain. All the hotel landscape details will follow the language of the main buildings. Manual plastering by STUCCO, arched doorways and windows, terra cotta pantile, iron art, logs and other extremely delicate techniques create authentic Mediterranean resort landscape.

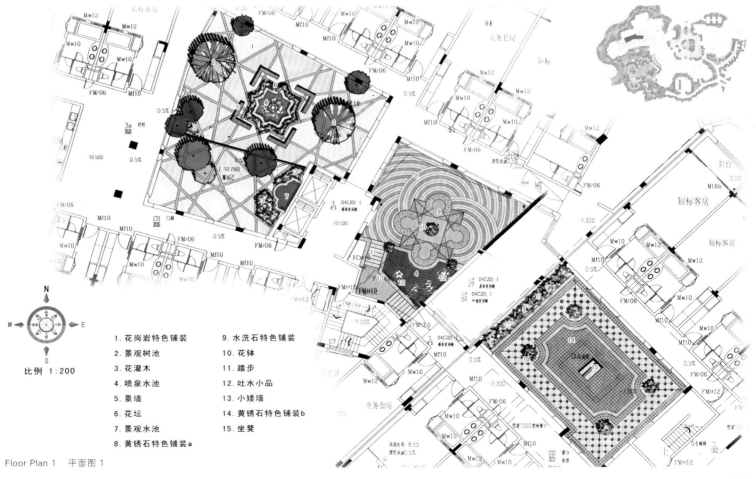

1. 花岗岩特色铺装
2. 景观树池
3. 花灌木
4. 喷泉水池
5. 景墙
6. 花坛
7. 景观水池
8. 黄锈石特色铺装a
9. 水洗石特色铺装
10. 花钵
11. 踏步
12. 吐水小品
13. 小矮墙
14. 黄锈石特色铺装b
15. 坐凳

比例 1:200

Floor Plan 1　平面图 1

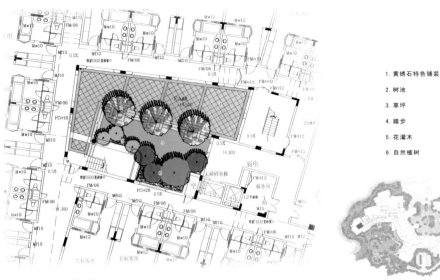

1. 黄锈石特色铺装
2. 树池
3. 草坪
4. 踏步
5. 花灌木
6. 自然植树

比例 1:200

Floor Plan 2　平面图 2

项目概况

美林湖温泉大酒店位于中国美林湖的核心区域——主题酒店园区，项目的自然条件优越，拥揽3万亩原生山林、湖泊、温泉、高尔夫，建筑主体依山面湖，以"意大利山城"为主题进行项目设计。

多个依山势而建的地中海风情温泉泳池呈跌级式排列，所有酒店景观的细节都遵循建筑主体的语言，以STUCCO手工抹墙、拱形门廊窗户、红陶筒瓦、铁艺、原木等极为考究的细腻手法，营造出原汁原味地中海式度假酒店景观。

Landscape Design

It uses stairs and low retaining wall approaches to skillfully handle height difference issue. Stairs combined with the murmuring falling water and low retaining wall rhythmically make everything so harmonious according to local conditions. Many Mediterranean style hot spring swimming pools are built down along the mountain. The fun sculptures and fountains above the pools add more vitality to peaceful environment.

Floor design uses red brick, ceramic mosaic and pebble mosaics, etc. which shows romantic and pleasant Mediterranean flavor. Another feature of this design is the application and combination of materials. In order to show the simplicity, design makes use of south gravel, and STUCCO paint, which make this hotel landscape better integrate into the environment in the entire forest.

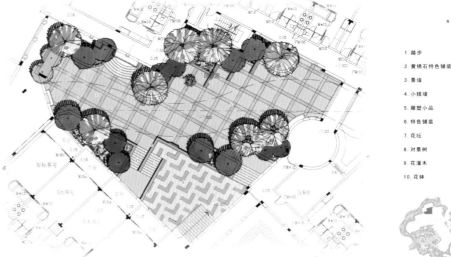

Floor Plan 3　平面图 3

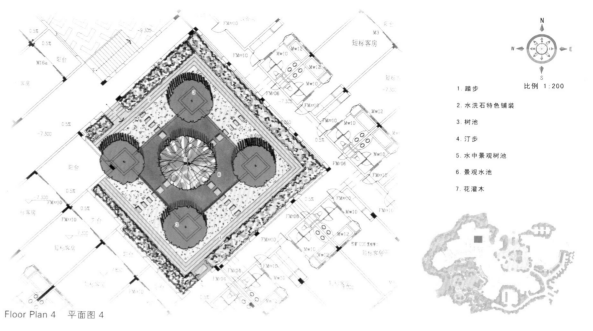

Floor Plan 4 平面图 4

Plant configuration uses tall palm plants to add momentum to the entrance; banana combines with wooden pavilion in the leisure place for stay; bougainvillea and concise STUCCO plastering combine to foil the blooming plants; climbing plants combine with pergolas and building walls to soften tough buildings. The moderately luxury landscape perfectly integrates acres of native hills with Mediterranean-style architectures, making it an ideal place full of vitality for relaxation and holiday.

景观设计

景观中利用台阶和挡土矮墙的手法娴熟地处理高差的问题，台阶结合潺潺的跌水，挡土矮墙有韵律地组合，这使得一切都是那么的和谐，因地制宜。多个依山势而建的地中海风情温泉泳池呈跌级式排列，泳池上的趣味雕塑和喷水为宁静的环境增添了更多的欢快。庭院中则用小雕塑水景和陶钵鲜花去营造宁静雅致的氛围。

铺地的设计中用了红砖、陶瓷拼花和卵石拼花图案等，这些无一不表现出浪漫宜人的地中海风情。本次设计的另一个特点是材料的应用和结合，为了表现淳朴的风情，设计中大量运用了南砂石、STUCCO 涂料等，这些都使这个酒店的景观更好地融入到整个山林的大环境中。

植物配置中用了高大的棕榈科植物为入口增添气势；在休闲停留的地方用芭蕉与木亭结合；勒杜鹃与简洁的 STUCCO 抹墙结合，更加衬托了植物的娇艳；攀爬植物则与花架和建筑墙体结合，柔和了建筑的硬朗。低调奢华的景观将万亩原生湖山与地中海风情建筑完美融合，使其成为了享受悠闲假期、焕发身心活力的理想之地。

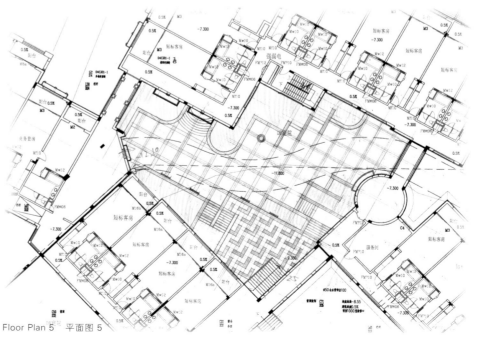

Floor Plan 5 平面图 5

Commercial Landscape
商业景观

KEYWORDS 关键词

Chinese Element
中式元素

Landscape Sketch
景观小品

Green & Ecology
绿色生态

Modern Minimalist Style
现代极简抽象风格

Location: Tai'an, Shandong
Landscape Design: Botao Landscape (Australia)
Landscape Area: 42,000 m²

项目地点：山东省泰安市
景观设计：澳大利亚·柏涛景观
景观面积：42 000 m²

Xinhua City International Plaza Commercial Area Landscape, Tai'an
泰安新华城国际广场商业区景观

FEATURES 项目亮点

The landscape design of the whole commercial area mainly embodies a green and ecological ambiance. It adopts modern Chinese element to form its own characters on the application of details, and uses lighting to strengthen the dynamic, atmospheric and decorative effect and to lead commercial flow and gather popularity.

整个商业区的景观设计，总体上体现绿色生态的氛围；在细节上运用现代中式元素形成自己的特色，通过灯光运用强化动感效果、气氛效果和装饰效果，引导商业人流，汇聚人气。

Overview

Located in the urban modern commercial area of Tai'an, Shandong, the project is positioned as a city landmark which integrates culture, tourism, leisure and artistic connotation and aims to build an urban complex commercial park with comprehensive functions. The overall architectural design is in modern concise and Chinese style. The landscape design adopts the approach with strong modern sense and certain Chinese cultural characters under the premise of maintaining the architectural color and style, and uses modern minimalist style to create landscape environment.

项目概况

项目位于山东省泰安市城市现代商贸区，被定位为集文化、旅游、休闲、艺术内涵为一体的城市地标，打造成具有复合功能的城市综合商业公园。整体的建筑设计风格为简约现代中式风格。景观设计在延续建筑的色彩和风格的前提下，采用现代感强烈，又具有一定的中式文化特色的设计手法，利用现代极简抽象风格来塑造景观环境。

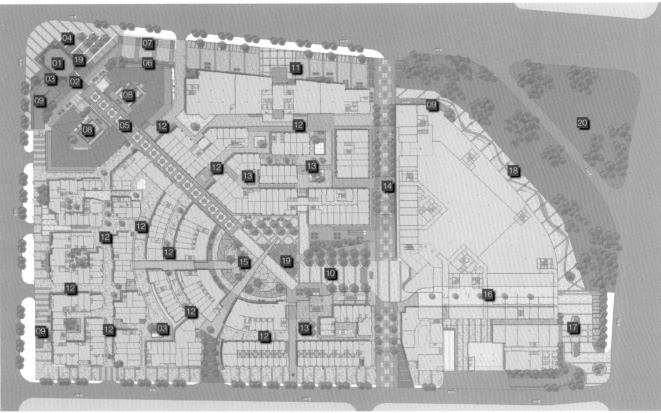

01	Main Plaza 主广场
02	Bridge 景桥
03	Element Pots(5 Elements) 元素花钵（5种元素）
04	Signage 标识
05	Lantern Parade 景观照明
06	Landscape Pond 景观池
07	Parking 停车区
08	Balcony 阳台
09	Basement Entry 地库入口
10	Central Plaza 中心广场
11	Activity Plaza 活动广场
12	Corridor 走廊
13	Small Plaza 小广场
14	Internal Driveway 内部车行道
15	Peony Plaza 牡丹花广场
16	Elevated Walkway 抬升走道
17	Hotel Drop-off 酒店落客
18	Mall Plaza 商业广场
19	Dry Fountain 旱喷
20	Civic Green 市政绿化

Site Plan 总平面图

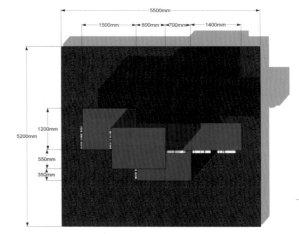

Floor Plan 平面图

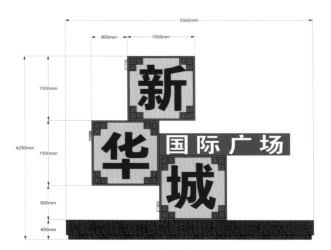

Elevation 立面图

Rendering 1 透视图1

Rendering 2 透视图2

Landscape Design

The landscape design of the whole commercial area mainly embodies a green and ecological ambiance. It adopts modern Chinese element to form its own characters on the application of details, and uses lighting to strengthen the dynamic, atmospheric and decorative effect and to lead commercial flow and gather popularity. Under the premise of maintaining the modern style, the designer integrates Chinese cultural ideas and concepts and divides the whole commercial area into five themes — gold, wood, water, fire and earth.

Gold — office and retail area: gold is used to imply the business prosperity. Various metal flowerpot lampposts and sketches are used to match the theme, forming a prosperous landscape business zone.

Wood — central plaza and the south area: it's a business leisure area, where wooden leisure terrace has been used to form a landscape space. The exterior and interior of the ecological business center transit naturally, and part of the area is embellished with unique and exquisite planting beds and flowerpots to form a rich visual effect and meet people's shopping and leisure demands.

Water — east commercial street: it's an open commercial space, and the floor is paved in the image of water to form a vivid and free ambiance.

Fire — main entrance at the northwest corner and the commercial corridor: the entrance plaza is concise and elegant, welcoming visitors worldwide. The commercial corridor uses lampposts and lanterns to form a warm ambiance.

Earth — southwest area: it's a small retail and catering area, where different colors and materials have been used to express the theme and form a warm and comfortable commercial street.

景观设计

整个商业区的景观设计，总体上体现绿色生态的氛围；在细节上运用现代中式元素形成自己的特色，通过灯光运用强化动感效果、气氛效果和装饰效果，引导商业人流，汇聚人气。在延续现代风格的前提下，将中式的文化思路和理念贯穿其中，将整个商业区划分为五个主题——金、木、水、火、土。

金——办公和零售区域：利用"金"来寓意商业繁荣，运用各种金属花钵灯柱及小品等契合主题，形成一个气氛浓烈的景观商业带。

木——中心广场及南区：商业的休闲区域，在这个区域里面，利用木质休闲平台来组织景观空间，生态商业中心室内室外自然过渡，局部点缀特色精致的种植池、花钵，形成丰富的视线效果，满足人群购物休闲需求。

水——东面商业街：商业开敞空间，地面铺装以水的形象展现，形成灵动自由的氛围。

火——西北角主入口和商业走廊：入口广场简洁大气，欢迎四方游客，商业走廊运用灯柱、灯笼构成热情的氛围。

土——西南片区：小型零售业及餐饮片区，利用不同色彩和材质来表达主题，形成一个温馨舒适的商业街区。

Frontage Elevation 1　沿街立面图 1

Frontage Elevation 2　沿街立面图 2

Frontage Elevation 3　沿街立面图 3

Frontage Elevation 4　沿街立面图 4

Commercial Landscape
商业景观

KEYWORDS 关键词

Spacious & Bright
宽敞明亮

Chongqing Characteristics
重庆特色

Natural & Relaxing
自然放松

Modernist Style
现代主义风格

Location: Chongqing
Landscape Design: ASPECT Studios
Land Area: 26,000 m²

项目地点：重庆市
景观设计：澳派景观设计
占地面积：26 000 m²

Chongqing Vanke Xijiu Plaza
重庆万科西九广场

FEATURES 项目亮点

The central "Stage" tree pool has been planted with three big trees whose branches stretch beautifully, making the space of the plaza more comfortable and pleasant, and providing people with natural sense of relaxation among the busy and noisy urban environment.

中央"舞台"树池种有三棵树形美丽舒展的大树，使广场更加舒适宜人，让人们在忙碌吵闹的都市环境中得到自然带来的放松感。

Overview

Vanke·Xijiu Plaza is located at the center of Shipingqiao retail hub and close to Yangjiaping business circle. It shares the convenient urban transportation system and complete surrounding facilities. Vanke Group has the vision of developing Xijiu Plaza into a vibrant urban space for the young generation to work, live and play.

项目概况

万科西九项目位于石坪桥核心位置，紧邻杨家坪商圈，尽享便捷城市交通体系，周边配套齐全。万科西九依托这一优势打造"杨家坪商圈一站式优活乐享地"，成为城市精英全新生活中心。

Development Idea

Formerly a west suburban hospital site with 50 years' history, it will be reconstructed with a commercial mixed-use development, comprising residence, commerce and business. The project has a land area of 26,000 m², and contains three high-rise residences, commercial street, apartment LOFT towers and restoration houses.

Planning Concept

One target of the design is to capture the local historical culture and natural topography and to inherit the local culture of Chongqing. The jumping lines of Chongqing mountains and the scene of two jointed rivers are reflected by artistic and abstract lines in forms of pavements, planting pools with benches, steps, drainage and so on to jointly create a semi-enclosed space. Elevated edge form of the site strengthens the sense of visual guidance and focuses people's view to the mall.

开发思路

场地的旧址是一个具有 50 年历史的西郊医院，被改造为一个集居住、商业、商务为一体的功能性的商业综合体。项目占地 2.6 万 m²，包括三栋高层住宅、商业街、公寓楼与还建房。

规划理念

本项目目标之一在于捕捉场地历史文化及自然风貌，传承与发扬重庆本土文化。重庆山脉跳跃的线条以及两江交汇的情景被一种艺术抽象的线条体现，形式包括铺地、带座椅收边的种植池、台阶、水系等，共同打造了一个半围合式的空间。场地抬高的收边形式强化视觉导入感，将行人的目光聚焦到商场。

Landscape Design

The lighting design of the plaza is diverse and bright, compromising punctate small spotlights hung in the sky of the plaza, projection lamps lighting up the fig tree, buried lights lighting up the dry fountains and lined hidden lights lighting up the edge of the bench tree pool; it also borrows the light of the surrounding street lamps.

The artistic features of Jiulongpo embody in the gathering space of the artists under the corridor and the stage for artworks exhibitions; at night, the "starry sky" artworks are exhibited in the sky above the ficus lacor, arousing the reunion scene of enjoying the cool Chongqing summer evenings at the courtyard. It remains part of the existing ficus lacor, and transplants some trees to provide shading for the site; the central "Stage" tree pool has planted three big trees with beautiful stretches, making the space of the plaza more comfortable and pleasant, providing people with natural sense of relaxation among the busy and noisy urban environment.

The plaza is quite spacious, and it can meet the needs for different activities and enables people of different ages, interests and cultural backgrounds to carry out various activities. Now, Vanke Xijiu Plaza becomes an urban public space with strong Chongqing characteristics and a pleasant, memorable and innovative space.

景观设计

广场的照明设计多样化而明亮，主要有挂在广场天空的点状小射灯、照亮无花果树的投射灯、照亮旱喷泉的地埋灯、照亮座椅树池边缘的线状藏灯，同时还借助了周围路灯的光辉。

九龙坡的艺术特征在于廊架下方艺术家的聚会空间以及艺术品展示舞台；在晚上，黄桷树的上空呈现"星空"艺术品，唤起重庆夏季夜晚院落纳凉团聚的景象。保留局部现有的黄桷树，并移植部分的树林从而为场地提供绿荫；中央"舞台"树池种有三棵树形美丽舒展的大树，使广场更加舒适宜人，让人们在忙碌吵闹的都市环境中得到自然带来的放松感。

商业广场非常开阔，可以满足不同活动的举办要求，同时也可以让不同年龄、不同兴趣、不同文化层次的人们开展多种活动。万科西九广场是一个带有重庆特色的都市公共空间，是一个欢乐、难以忘怀的创新空间。

Commercial Landscape
商业景观

KEYWORDS 关键词

High Mountain and Flowing Water 高山流水

Diversified landscape 多元景观

Surrounded by Mountain and Water 山环水抱

New Chinese Style 新中式风格

Location: Haidian District, Beijing
Client: Beijing Haixin Ark Real Estate Development Co., Ltd.
Landscape Design: America Leedscape Planning and Design Co., Ltd.
Landscape Area: 160,000 m²

项目地点：北京市海淀区
委托单位：北京海欣方舟房地产开发有限公司
景观设计：美国俪禾景观规划设计有限公司
景观面积：160 000 m²

Dongshenghui
东升汇

FEATURES 项目亮点

The project has introduced traditional and classical landscapes into modern leisure spaces to create unique spatial and visual experience. It has also well blended Western culture into Chinese culture by skillful design.

项目将传统的古典园林融入现代休闲空间，营造出独特的空间氛围与视觉感受；加之设计中静动结合，使中西方文化完美结合，成为设计中的典范。

Overview

The rectangular site locates within Xixiaokou green zone, Haidian District of Beijing, occupying an area of 160,000 m², of which, the construction area will be 10,000 m² and the existing water surface is 23,000 m². It is conceived as a high-end, mixed-use community that integrates office, culture, entertainment, education, healthcare, and Eco residence, to realize people's dream of living with nature. By building a place with high mountain and flowing water, it will "free" people from the "cage" of steel and concrete, and enable them to go back to nature.

项目概况

该地块位于北京市海淀区西小口绿化隔离区内，规划占地面积约16万m²，其中规划建筑占地约1万m²，现有水面2.3万m²。地块呈长方形。项目定位为高端复合社区，将商务办公、文化娱乐、教育医疗、生态居住融为一体，满足久居城市的人们对大自然有着发自内心的向往。通过构建这一处"高山流水"的清悠，满足那"久在樊笼中，复得返自然"的渴求。

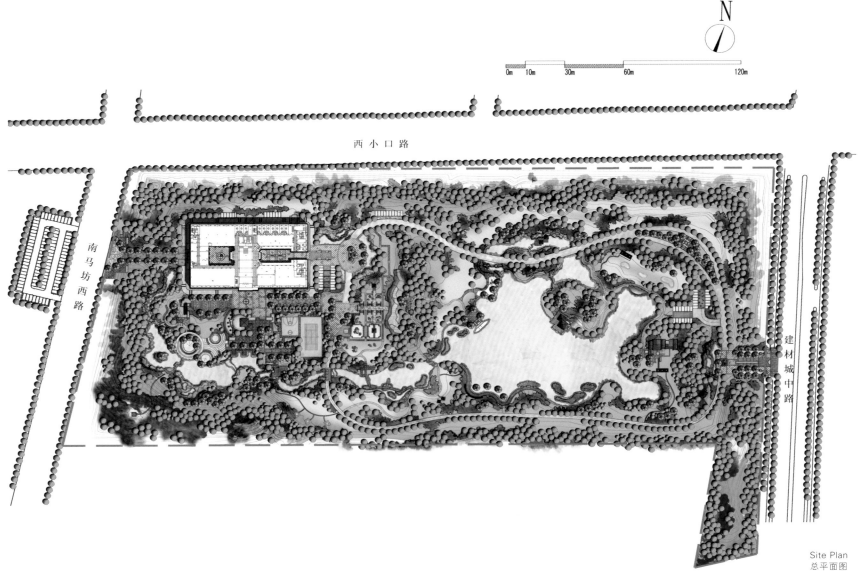

Site Plan
总平面图

Landscape Design

The site itself has already outlined the framework of the landscape system, and the "high mountain", "large lake" and "streams" are natural. So, based on the principle of "sustainable development and economical construction", the design has taken advantages of the site and reserved the framework, establishing a diversified landscape system by cleverly dealing with the details and adding new landscape elements.

The project has introduced traditional classical landscapes into modern leisure spaces to create unique spatial and visual experience. It has also well blended Western culture into Chinese culture by skillful design. Moreover, the landscape design is combined with architectural design to emphasize the beauty of harmony.

The design has introduced the theory of "Feng Shui", a Chinese philosophical system of harmonizing everyone with the surrounding environment, into the project, trying to analyze the characteristics of the universe, earth and humanity, and building a harmonious environment that promotes the development of all of them. Keeping this theory in mind, the designers have created a favorable landscape environment.

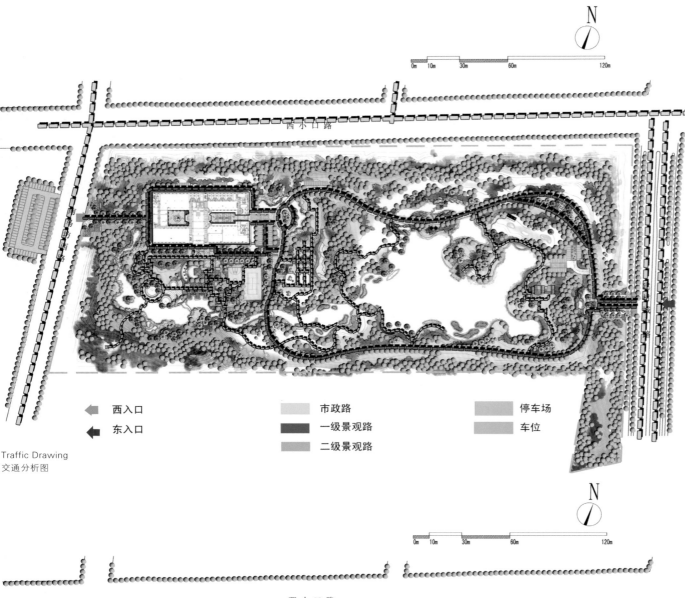

Traffic Drawing
交通分析图

景观设计

基地大体的山水骨架已然初成，且"高山"—"大湖"—"溪流"之势似浑然天成，如墨游动，动如行龙。所谓"文章本天成，妙手偶得之"，故本着适度改造、经济建设原则，方案采取随形就势、因地制宜的设计手法，保留大体山水骨架，佳则守之，劣则改之抑或屏之，不足之处补之，有过之处减之，增加细节，巧妙构思，融入新的景观元素，形成多元景观。

项目将传统的古典园林融入现代休闲空间，营造出独特的空间氛围与视觉感受；加之设计中静动结合，使中西方文化完美结合，成为设计中的典范；景观设计与建筑设计相得益彰，相互融合，相互衬托形成了和谐之美。

天地人神的和谐统一。天地人神和谐统一的系统规律是风水理论的本质。具体而言，就是综合分析天地人所在的五行属性及该五行属性的运动规律，在此基础上，通过对人及其所处环境的优化设计布局，创造出有利于人顺畅发展的吉祥气场，从而达到天地人的和谐统一。引入风水理念，演绎背山面水、山环水抱、风生水起、藏风聚气的景观形态。

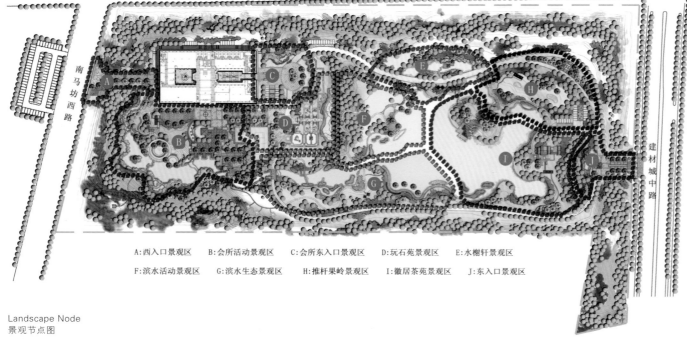

Landscape Node
景观节点图

East Entrance Section
东入口剖面

Club Area Section
会所区剖面图

Commercial Landscape
商业景观

KEYWORDS 关键词

Leisurely Style
休闲风情

Ecology and Art
生态艺术

Landscape Features
景观小品

Modernism Style
现代主义风格

Location: Guangzhou, Guangdong
Developer: China Vanke Co., Ltd. / Guangzhou Vanke Real Estate Co., Ltd.
Landscape Design: PASNO Landscape Architecture Studio

项目地点：广东省广州市
开发单位：万科集团 / 广州市万科房地产有限公司
景观设计：普梵思洛（亚洲）景观规划设计事务所

Vanke Northern Dream Town, Guangzhou
万科广州北部万科城

FEATURES 项目亮点

By adhering to the overall design and carefully choosing materials, the pavement is designed to be simple, modern, stylish and elegant, forming sexy textural effect.

铺装设计讲求简约，洋溢着现代、时尚、干练的气质，铺装设计遵循整体的设计理念，从铺装的性感肌理到选材意向都贯彻了这种风格。

Overview

Located in Prince Hill ecological zone, at the border of Guangzhou Huadu District and Qingyuan City, Vanke Northern Dream Town is one of the key projects in the "northern city development strategy" of Guangzhou with great potential. Boasting a total floor area of approximately 1,300,000 m², it is conceived as a large-scale and ecological urban community accommodating a population of about 40,000 to 60,000. It includes five big areas, namely, the park area, the education area, the commercial area, the F&B area and the residential area, providing complete supporting facilities, such as the traffic island, the supermarket, the food street, the high-quality primary and middle schools, the community hospital as well as the lakeside club, KTV, gym, SPA center, mountain sports park, etc.

项目概况

项目位处广州花都区与清远交界处的王子山生态区，是广州城市发展规划中"北优"战略的核心开发区域，发展前景广阔。项目总建筑面积约为 130 万 m²，未来将形成约 4～6 万居住人口规模的大型生态型城市社区。项目超大规模自成一体，内部分为公园区、教育区、商业区、餐饮区和住宅区共 5 大区域，完整的生活设施体系规划，包括社区交通岛、生活超市、餐饮街、优质中小学、社区医院以及临湖会所、KTV、运动馆、SPA 水疗中心、山体运动公园等休闲设施。

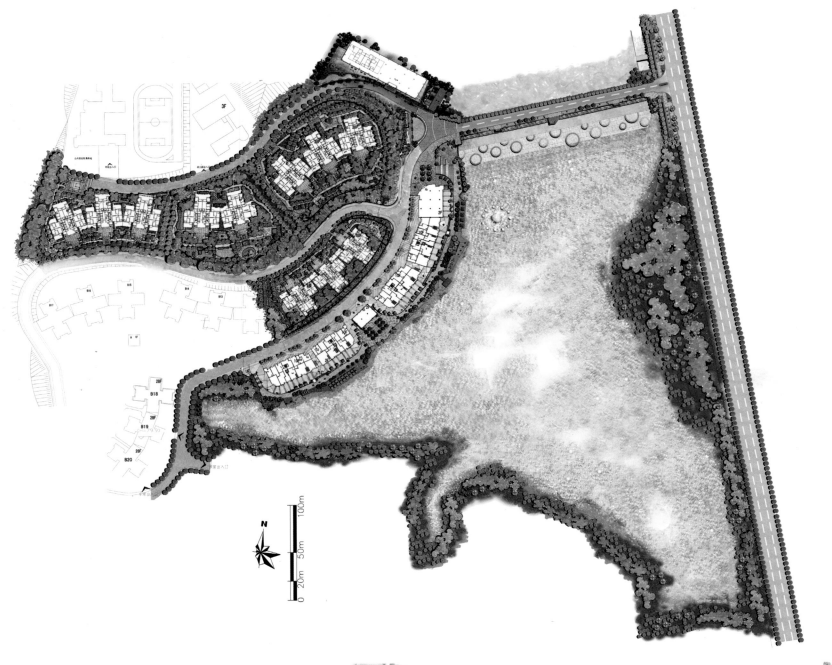

Site Plan
总平面图

Design Idea

The designers proposed that the design should well deal with the relationship between man and nature as well as man and environment. Natural elements include water (lake, waterfall, cascade, shallow water and mirror-like water); stone (the use and recombination of local stones); dam (break the boundary between man and dam, open the spaces beside the dam, establish a vertical landscape system, and encourage people to get close to the dam); trees (show their functions to offer shades and adjust air humidity). It tries to create a new comfortable environment which allows people to get close to nature. The aim of construction is to introduce healthy and friendly elements by making reference to the theory of Defeated Architecture.

设计理念

设计师提出：在设计上需要解决的问题就是：人与自然的关系、人与环境的关系：水（湖水、瀑布、跌水、浅水、镜面水）、石（当地石材的运用与重组）、坝（解决"人与坝"冷酷的隔绝关系，打通入口处大坝两边的空间；丰富立体景观提高人的参与机会）；树（尽显其天然的遮阴与调节湿度功能）；人（提供舒适的新环境，让人与自然亲密接触）；构筑（借鉴"负建筑"的原理，引进健康并与人亲近的元素）。

Design Skills

The designers have defined the project in modern natural and riverside leisurely style, trying to create the following landscape features:

1. Main Entrance and Dam: the entrance space is natural and peaceful with exuberant vigor and unique personality. The dam is simple but elegant which will lead people back to their home directly.

2. Opposite Views: unpretentious water screen and pleasant sunken lawn reinterpret the new experience of modern nature to people.

3. Commercial Square: jumping fountains, swaying trees, funning sculptures and interesting signs are well organized to enhance the dynamic atmosphere of the commercial square.

Lakeside Commercial Street: colorful lighting, leisure bar, lively ripples and the multi-functional service area help to create a colorful lifestyle.

Phase II tries to present simple and elegant landscape of modernism style, with emphasis on abstraction and order, simplicity and unpretentiousness, ecology and art, science-technology and quality. By adhering to the overall design and carefully choosing materials, the pavement is designed to be simple, modern, stylish and elegant, forming sexy textures. Simple and elegant lights and lamps are chosen and installed reasonably to save energy, guide people effectively and highlighting the beautiful night views of the key landscape nodes.

设计手法

项目设计风格为现代自然＋临江休闲风情，设计师力求打造以下景观亮点：

1. 主入口及大坝：入口空间表达自如，平淡中见奇掘、轻落中有纵生，仿佛有一种力量在心头跳跃闪动；夸张被隐藏，浮华被撕去，个性化的情愫被体现。大坝简约而不简单，放松空间、长远的视线，直接将人引入温馨的家园。

2. 对景：含蓄的水幕、宜人的下沉草坪，将现代自然的全新感受向人们重新诠释。

3. 商业广场：跳动的旱喷、婆娑的树阵、嬉戏的情景雕塑、趣味的指示牌将商业广场的氛围生动的组织起来。

4. 湖滨商业街：灿烂的灯光、休闲的酒吧、灵动的水波、多样的休闲服务区，为人们提供了丰富的生活情趣。

二期组团力求打造简约的现代主义社区景观，讲求抽象与秩序、质朴与内敛、生态与艺术、科技与品质。铺装设计讲求简约，洋溢着现代、时尚、干练的气质，铺装设计遵循整体的设计理念，从铺装的性感肌理到选材意向都贯彻了这种风格。灯具选型也以简洁为主，以环保、节能为设计理念，进行合理的照度控制和布点设计，突出安全、引导功能和重要景观节点的夜景效果。

Commercial Landscape
商业景观

KEYWORDS 关键词

Ecological Green Island
生态绿岛

Green Corridor
绿色廊道

Full of Vigor and Vitality
生机盎然

Spanish Style
西班牙风格

Location: Suqian, Jiangsu
Developer: Suqian Powerlong Property Development Co.,Ltd
Landscape Design: Botao Landscape (Australia)
Landscape Area: 62,000 m²

项目地点：江苏省宿迁市
开发商：宿迁宝龙置业发展有限公司
景观设计：澳大利亚·柏涛景观
景观面积：62 000 m²

Powerlong Plaza, Suqian, Jiangsu
江苏宿迁宝龙广场

FEATURES 项目亮点

The leisure waterfront zone is composed of "Barcelona ecological island" and "Bilbao riverside leisure zone" which, with emphasis on natural-style landscape design and planting, have created a fascinating and yearning green space for people.

整个休闲滨水带由"巴塞罗那生态岛"和"毕尔巴鄂滨河休闲带"两部分组成。这两部分在景观设计和植物种植上注重因地制宜，共同打造了一个让人陶醉、向往的绿色场所。

Overview

Located on the West Lake Road of Suqian City, Jiangsu Province, Powerlong Plaza is conceived as one of the top developments in North Jiangsu for shopping, recreation, entertainment and living.

项目概况

宿迁宝龙城市广场项目位于江苏宿迁市西湖路，项目的建设目标是成为苏北地区首屈一指的商业购物、休闲娱乐中心和生活社区。

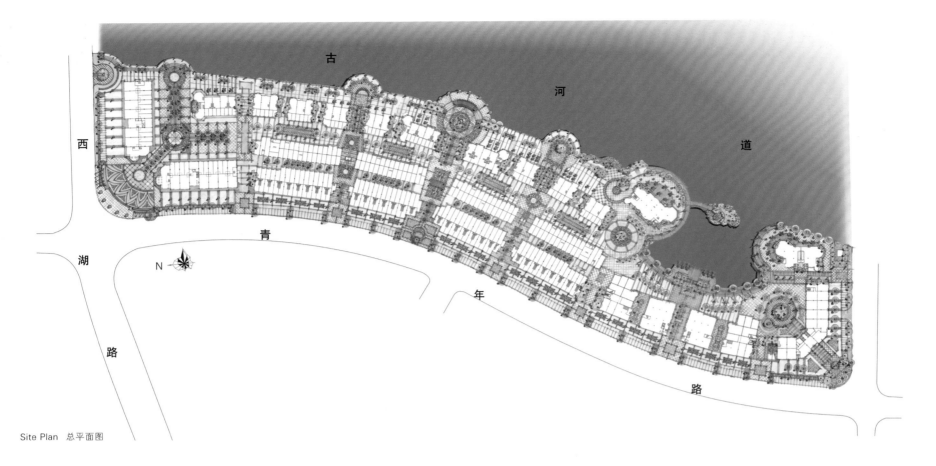

Site Plan 总平面图

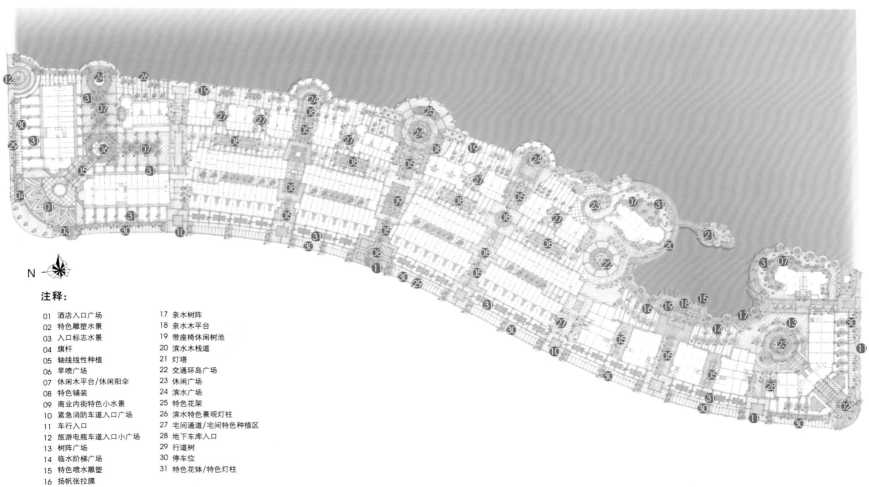

注释：

01 酒店入口广场
02 特色雕塑水景
03 入口标志水景
04 旗杆
05 轴线线性种植
06 旱喷广场
07 休闲木平台/休闲阳伞
08 特色铺装
09 商业内街特色小水景
10 紧急消防车道入口广场
11 车行入口
12 旅游电瓶车道入口小广场
13 树阵广场
14 临水阶梯广场
15 特色喷水雕塑
16 扬帆张拉膜
17 亲水树阵
18 亲水木平台
19 带座椅休闲树池
20 滨水木栈道
21 灯塔
22 交通环岛广场
23 休闲广场
24 滨水广场
25 特色花架
26 滨水特色景观灯柱
27 宅间通道/宅间特色种植区
28 地下车库入口
29 行道树
30 停车位
31 特色花钵/特色灯柱

Floor Plan 平面图

Ecological Island Landscape

The Barcelona Eco island is surrounded by buildings on three sides to present a pleasant and friendly bow-shaped space. Comfortable coastal square, vivid fountain sculpture and the soaring green trees make the square full of vigor and vitality. The eye-catching white tower, standing confidently on this Eco island, seems always to be the focus of the view.

生态岛景观

巴塞罗那生态岛区，由三面建筑群和一个生态岛围合而成的"弓"形空间，亲切宜人。温馨舒适的沿岸广场、栩栩如生的喷水雕塑、挺拔高歌的遮荫绿树增加了广场的透视效果，使广场显得更加生动、深远。醒目的白塔，自信地矗立在生态岛上，好像永远都是镜头的焦点。

Slope Section B1　坡面图 B1

Landscape Design

Taking advantage of the strategical location, with emphasis on the relationship between human and nature, it aims to create a modern commercial block of Spanish style by the lake. The landscape system consists of "one belt, three interfaces and four axes" — "one belt" refers to the leisurely waterfront area, "three interfaces" includes two entertainment and business areas and one commercial street, and the "four axes" well connects the east with the west.

The "belt" covers the Eco island, the leisurely waterfront revetment, the sightseeing route for battery cars and the waterfront commerce. It is a dominant and ideal place where people yearn for. And, according to the municipal requirements, a sightseeing route for the battery cars should be set along the revetment. Thus the designers have fully considered the landscape, commercial and social effect of this area, and shown respect to local culture to optimize this belt with people-oriented designs.

Buildings of "three interfaces and four axes" are mainly designed in European style with emphasis on landscape nodes at the beginning, middle and end of the axes. European-style landscape elements are used to interpret the characteristics of the Spanish-style commercial area and present romantic, natural and elegant landscape spaces. At the same time, it pays attention to the functions of the landscape, trying to build a stylish European street for sightseeing, shopping and entertainment.

景观设计

景观设计充分利用滨河的优越地理位置，注重人与自然的融合，旨在打造具有现代气息的西班牙风情滨河商业街区。景观理念为"一带三面四轴线"："一带"指休闲滨水带；"三面"指两个娱乐商务区加一个风情商业街区；"四轴"指衔接东西向的四个主轴。

"一带"是由生态岛、休闲滨水驳岸、电瓶观光道及东面临水商业圈构成。在整个地块中占主导地位，也是最让人陶醉、向往的好去处。应市政府要求沿河堤设一条电瓶观光道，设计师充分考虑这"一带"的景观设计所带来的景观标志性效应、商业消费效应及社会效应，同时，尊重地域人文，尊重人本设计，使这"一带"在景观与观景中得到升华。

"三面四轴"以欧式建筑风格为主调，强调轴线与衔接的"头、中、尾"标识性景观节点，借助欧式景观元素来诠释西班牙浓烈风情的商业圈，体现浪漫高雅、自然大方的景观空间，同时，慎重考虑景观的经济实用性，力求打造集"旅游观光、购物消费、娱乐休闲"一体化的风情欧洲街。

Leisure Zone Landscape

Bilbao riverside leisure zone, with fantastic scenery and a distance from hubbub, presents a new world that connects the river culture, important social contact and catering culture. It is also a potential tache for tourism development. Several leisurely squares run throughout this zone to create a vibrant and artistic atmosphere, allowing tourists to shop and eat delicious food while enjoying the beautiful views.

Alicante Square is the holy place for matadors. The magnificent bullfight gates in the square use architectural elements of Roman amphitheatre and are painted in bright color to echo the red cloaks of the matadors.

休闲带景观

毕尔巴鄂滨河休闲带，风光旖旎，远离尘嚣，是一条连接河畔文化、重要社交、饮食文化的新天地，还是发展旅游业潜力的纽带。贯穿其中的几个休闲广场，时刻焕发活力，处处彰显艺术氛围，游客在惬意游玩的同时，还可以伴着美景尽享购物和美味佳肴。

阿里冈特广场是斗牛士们的"圣地"，广场上气派的斗牛门，提炼古罗马剧场式的圆形建筑，色彩明快，仿佛是在呼应斗牛士那夺目的红披风，品味卓越，美轮美奂。

Plant Design

In terms of plant design, the Eco island is mainly covered by natural-style planting. Plants of different heights are well combined to make the white tower embraced by lush green. Thus the plants growing on the water bank will form an ecological green island together with the aquatic plants.

Surrounding the Eco island are tall trees and neat shrubs which ensure an open views of the tourists and make the white tower be the visual focus of this area. Along the river bank is a dynamic tourism area which is an ideal place for walking and enjoying the beautiful river views. Therefore, tall trees with big canopies and well-trimmed shrubs have been chosen to provide dense shades and form a green corridor along the river.

植物设计

在植物设计方面，生态岛上的绿化以自然种植为主，植物高、中、低相互搭配，使白塔被簇拥于一片生机盎然的绿色当中，使水岸生长的植物与水生植物一起，形成一个自然、野趣十足的生态绿岛。生态岛周围的绿化以高大乔木加整齐灌木的形式布置，保证游人视线通透，让白塔成为该区域的视觉焦点。河堤一线为动态游览区域，是电瓶车观光和游人购物之余漫步、欣赏河道美景的主要路线。因此绿化应提供相应的遮阴，选用树形舒展、树荫浓郁的乔木和可修剪灌木搭配，形成河岸一条绿色的廊道。

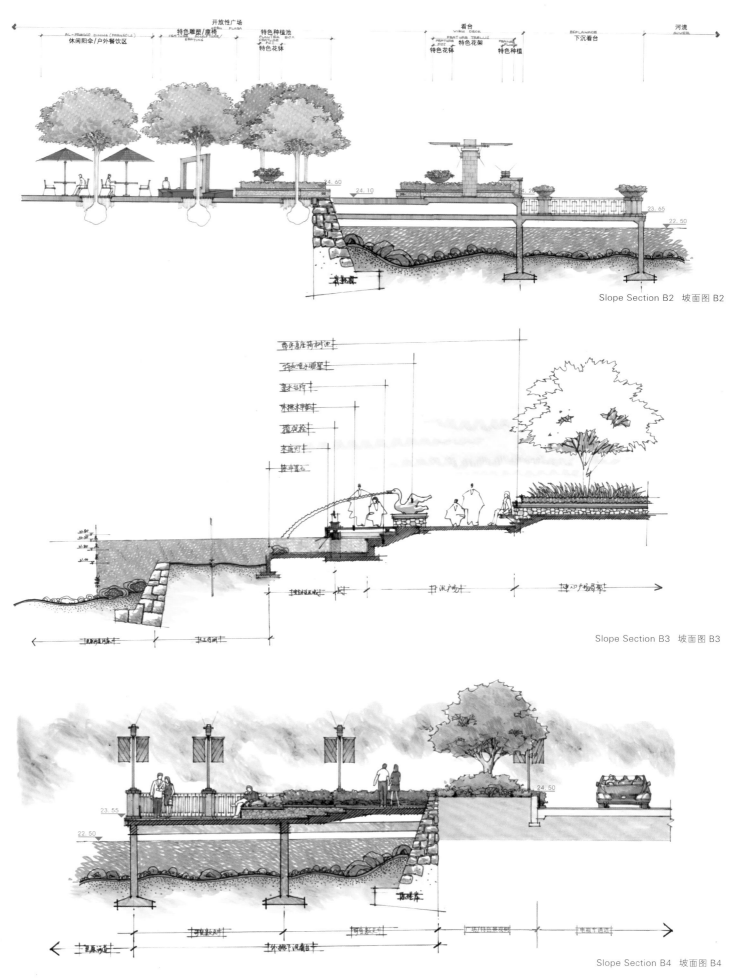

Slope Section B2 坡面图 B2

Slope Section B3 坡面图 B3

Slope Section B4 坡面图 B4

Commercial Landscape
商业景观

KEYWORDS 关键词

Natural Ecology
自然生态

Landscape Axis
景观轴线

Split-level
高差层次

Southeast Asian Style
东南亚风格

Location: Huizhou, Guangdong
Developer: Property Management Office of the Financial Street
Landscape Design: Siteline Environment Design Ltd.
Designer: Huang Jianfeng
Land Area: 36,498 m²
Floor Area: 71,500 m²
Green Coverage Ratio: 41%
Plot Ratio: 1.77

项目地点：广东省惠州市
开发单位：金融街置业
景观设计：SED 新西林景观国际
设计师：黄剑锋
占地面积：36 498 m²
建筑面积：71 500 m²
绿化率：41%
容积率：1.77

Haishang Bay Holiday Apartments, Financial Street of Xunliao Bay, Huizhou

金融街惠州巽寮湾海尚湾畔度假公寓

FEATURES 项目亮点

The project uses the "Andaman Coast" as design theme. While striving to build a landscape with sophisticated functions and amazing environment, the designers also obey the green and environment-friendly natural idea, trying to build the first landscape project that focuses on ecology and nature at the bay of Huizhou.

项目以"安达曼海岸"为设计主题。在全力打造功能完善、环境惊艳景观的同时，设计师也遵循了绿色、环保的自然理念，力求在惠州这段海湾上打造出重生态、重自然的景观设计先例。

Overview

Huizhou Haishang Bay is located at the Xunliao Bay Seaside Tourist Resort area, in the southern part of Huidong County, Guangdong Province, and is in the west of Renping Peninsula and on the east coast of Daya Bay. It boasts 982 units of apartments with each 70 m² in average, and a plot ratio of 1.77. The target customers are those longing for coastal life and those in need of investment.

项目概况

惠州海尚湾畔位于广东省惠东县南部巽寮湾海滨旅游度假区，稔平半岛西部，大亚湾东海岸。

公寓总套数为982，套均70 m²，容积率1.77。目标客户为向往滨海生活，有投资需求的消费者。

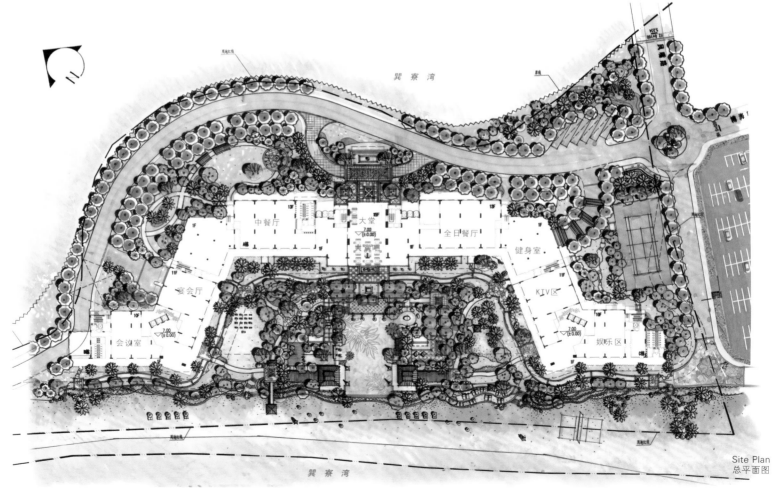

Site Plan
总平面图

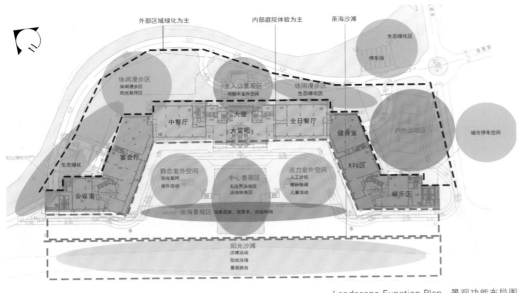

Landscape Function Plan 景观功能布局图

Landscape Axis Analysis 景观轴线分析图

Positioning Strategy

The project strives to be quality-based, and aims at building an international seaside tourist resort with high grade to make vacation, self-cultivation and the nature mutually integrated. Comparing to simple style, it is easier to create a stylish and paradise-like holiday experience by using tropical-style details.

定位策略

项目力求以质量为主打，追求高档次打造旅游度假、修养身心、与自然相互融合的国际化海滨旅游度假区。景观设计运用热带风格较强的设计细节语言，对比简约风格，更易营造一种风情化的、热带天堂般的度假感受。

Planning

The project is with rich Southeast Asian Style, and it borrows the original resource — the plot that is divided into two parts by Xunliao Bay, and it uses the Bohai Bridge as belt, playing a continuous leading role.

规划布局

项目建筑颇具有东南亚的热带风情，借助原有资源——被巽寮湾一分为二的项目地块，以博海桥为纽带，发挥着连续、引导性的作用。

Design Idea

The project uses the "Andaman Coast" as design theme. While striving to build a landscape with sophisticated functions and amazing environment, the designers also obey the green and environment-friendly natural idea, trying to build the first landscape project that focuses on ecology and nature at the bay of Huizhou. Among the whole project, various designs are servicing for the purpose of environmental protection. The local climate is similar to the Andaman Coast, which is easier for the designers to create similar landscape feelings from plants collocation, material utilization and so on. On lighting, waste classification and so on, the designers keep the idea of being natural, green and ecological, having effectively realized the ecological and sustainable development of the site's environment.

设计理念

项目以"安达曼海岸"为设计主题。在全力打造功能完善、环境惊艳景观的同时，设计师也遵循了绿色、环保的自然理念，力求在惠州这段海湾上打造出重生态、重自然的景观设计先例。整个项目中，多处设计都是为环保这个宗旨服务，当地气候条件上与安达曼海岸的相似性，更利于设计师从植物配置、材料运用等方面营造与其相同的景观感受。在景观照明、垃圾分类等各方面，设计师力求自然、绿色、生态，有效实现了场地环境的环保及可持续发展。

Section of Main Entrance 1　主入口剖面图1

Section of Main Entrance 2　主入口剖面图2

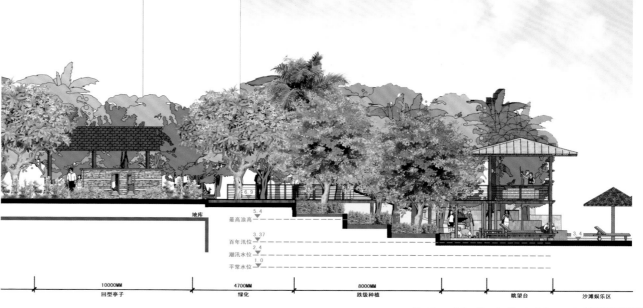

Section of Pavilion in Central Area　中心区亭子剖面图

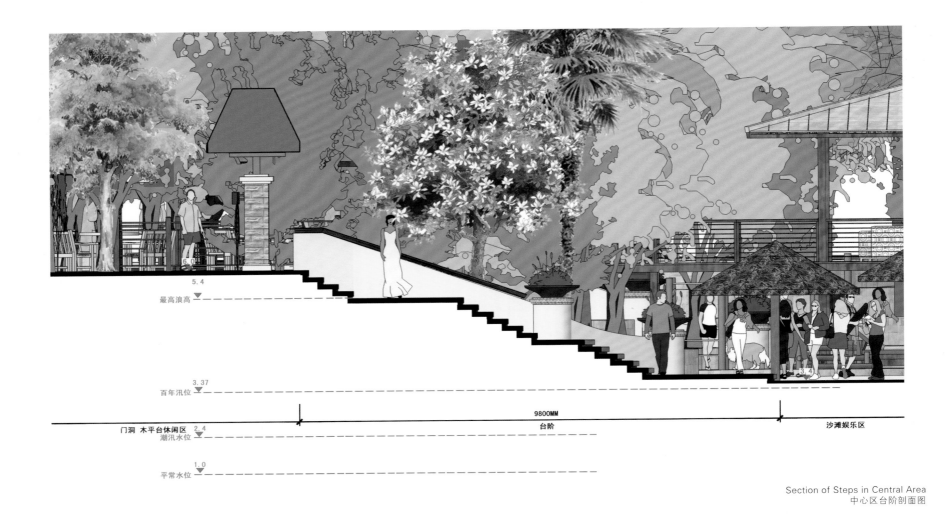

Section of Steps in Central Area
中心区台阶剖面图

Elevation of Beach in Central Area
中心区海滩立面图

Landscape Design

Combing with the road elevation difference, the main entrance area of the hotel is designed with landscapes of split-level sense. The surrounding is mainly with shrub landscaping, which better expresses the identity of the hotel and becomes the turning sight that guides the traffic flow.

Passing through the lobby and the water bar, the one comes into sight is the seaside tourist resort which is quite similar to the Andaman Coast and with coconut grove, tree shadows, small stones and white sands. The entire central landscape area is greening-based and with the utilization of natural materials and the design of Southeast Asian Style. The designers make the single pitched water bar landscape

景观设计

结合道路高差关系，酒店主入口区域设计出有高差层次感的入口景观。周围以灌木造景为主，更好地体现了酒店的标识性，形成引导车流的转弯视线。

穿过大堂水吧，出现在眼前的是与安达曼海岸极其相似的椰林树影、细石白沙的滨海度假区。整个中心景观区以绿化为主，加以自然材料的应用和东南亚风格构筑物的设计。以景观中轴的单坡水吧景观构筑物及无边界泳池为主要景观，延伸出两个休闲活动场地。这里设置了若干休息椅、遮阳伞及淋浴间，供客人休

structures on the landscape axis and the borderless swimming pool as main landscapes and then extend two leisure activity places where are arranged with several resting chairs, beach umbrellas and shower cubicles for guests to rest and chat. Along with the swimming pool and leisure places are several landscape walls, sculptures and split-level plants, creating an orderly enclosed and semi-private landscape space. The northwest area of the swimming pool is mainly designed with plants and the southeast area is arranged with kid's zone and enclosed corridor, and they use garage range lines as boundary to create a series of split-level landscapes. The humanized soft and hard collocations endow people with a heartfelt sense of belonging when first come here.

息和交谈。沿泳池与休闲活动场地，有多处景墙、雕塑、植物高低错落，营造出一个围合有序的半私密观空间。泳池的西北区域以植物设计为主，东南区设置儿童区和回型廊架，以地库范围线为界，打造一系列的落差景观。贴心的软硬配置考虑，让人们一来到这就有一种归属感由心而生。

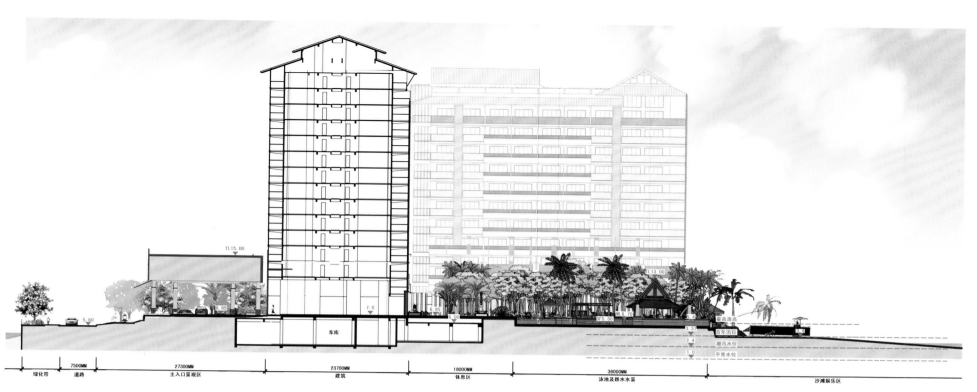

Section of Landscape in Central Area
中心区景观剖面图

Commercial Landscape
商业景观

KEYWORDS 关键词

Natural Harmony
自然和谐

Ecological Function
生态机能

Landscape Node
景观节点

Modernism Style
现代主义风格

Location: Huangdao District, Qingdao, Shandong
Developer: Shandong Xiaozhu Mountain Construction and Development Co., Ltd.
Landscape Design: Ecoland
Designer: David Chen
Land Area(Phase II): 27 mu
Landscape Area(Phase II): 9.5 mu
Green Coverage Rate: 50%
Plot Ratio: 1.0

项目地点：山东省青岛市黄岛区
开发单位：山东小珠山建设发展有限公司
景观设计：ECOLAND 易兰规划设计院
设计师：陈跃中
占地面积（二期）：27 亩
景观面积（二期）：9.5 亩
绿化率：50%
容积率：1.0

Vanke Tsingtao Pearl
万科青岛小镇

FEATURES 项目亮点

Based on the environmental background of "innumerable mountains and valleys, luxuriantly green trees and grass, and colorful clouds", the designers follow the rules of nature, exploit the favorable conditions and keep the most beautiful and natural elements of the site, trying to meet the functional requirements and achieve the harmony between nature and human's life.

"千岩竞秀，万壑争流，草木蒙笼，若云兴霞蔚"，在这样良好的基底环境中，设计师法自然，充分挖掘有利条件，将基地中最美、最真的一面保留下来，糅合适当的功能需求，达到与自然、与生活的和谐统一。

Overview

Tsingtao Pearl is located on Binhai Avenue of the west coast, at the southern foot of Xiaozhu Mountain. It is to the southeast of Qingdao Zhushan National Forest Park, about 1.6 km to the northeast of Tangdao Bay, about 5km away from downtown Huangdao, and about 10 km from the center of Jiaonan City. With the opening of the undersea tunnel and the cross-sea bridge, the west coast area is connected with the urban area of Qingdao City.

项目概况

青岛小镇位于西海岸滨海大道小珠山南麓。其西北与青岛珠山国家森林公园相邻，东南距唐岛湾海边约 1.6 km，距黄岛区市中心约 5 km，距胶南市中心约 10 km。随着海底隧道、跨海大桥的开通，西海岸区和青岛市区联为一体。

Master Plan
总平面图

Planning and Layout

The community will be developed in eight phases, covering a total planning floor area of nearly 1,000,000 m². Product types cover detached houses, semi-detached houses, townhouses, folding houses, apartments and other high-end residential products. Phase I covers a floor area of 79,000 m², including detached and semi-detached high-end houses, which locates in the east of Mid-Levels Area and is perfectly surrounded by mountains on three sides. Phase II is located in the southwest, with a total floor area of about 160,000 m². The entire project is designed and planned by dozen of internationally renowned teams. There are over 60,000 m² of international supporting such as high-end hotels, entertainments, educations, health and the world top resort Banyan Tree Hotel in the community.

Commercial Street was completed in advance. Ecoland takes charge of the overall landscape design in Phase II and focuses on planning the town's commercial street, villas and mountain parks.

规划布局

社区将分八期开发，总规划建筑面积近100万m²。产品类型涵盖独栋、双拼、联排、叠拼、公寓等高端居住产品。一期建筑面积约7.9万m²，主要为独栋、双拼高档别墅。整个一期位于半山区东部，踞于三山环抱的绝佳位置中。二期位于西南端，总建筑面积约16万m²。整个项目由十余家国际知名团队规划设计，小区内部辅以高端酒店、休闲娱乐、教育、养生社区等6万余平方米国际配套以及世界顶级度假酒店悦榕庄酒店作为配套。

商业街部分已于先期建成。项目二期由易兰进行整体景观设计工作，对小镇的商业街、别墅区及山地公园等部分重点规划。

Design Concept

Considering the special geographical conditions, Ecoland's designers have tried to break the traditional way of commercial design, using simple slates, rough and natural surface materials and swaying ornamental grasses to create a commercial street which is seemly born in mountains.

Landscape Design

The overall landscape structure of the town makes use of the building's height difference to display the landscape of miniature Xiaozhu Mountain in landscape design. In addition, it is worth mentioning that the landscape design combines a variety of modern design methods to make the whole town keep the plain temperament without losing the sense of the times.

Design has to respect nature, and makes it into life, because natural scenery is often intoxicating enough and nature itself is the best design paradigm. Based on the environmental background of "innumerable mountains and valleys, luxuriantly green trees and grass, and colorful clouds", the designers follow the rules of nature, exploit the favorable conditions and keep the most beautiful and natural elements of the site, trying to meet the functional requirements and achieve the harmony between nature and human's life.

设计理念

出于对其特殊地理条件的考虑，易兰的设计师力图打破传统的商业设计方式，利用质朴的板岩、粗矿的自然面材料以及在风中摇曳的观赏草，使得整个商业街营造出如同从山中孕育出来的观感。

景观设计

小镇整体的景观结构是利用建筑本身的高低落差，将小珠山的山水微缩于景观设计之中。此外值得一提的是，因为景观设计糅合了各种现代的设计手法，使整个小镇在保持质朴气质的同时又不失时代感。

设计须尊重自然，并使其融于生活，因为自然所呈现的景色往往足以令人陶醉，自然本身也就是最好的设计范式。"千岩竞秀，万壑争流，草木蒙笼，若云兴霞蔚"，在这样良好的基底环境中，设计师法自然，充分挖掘有利条件，将基地中最美、最真的一面保留下来，糅合适当的功能需求，达到与自然、与生活的和谐统一。

Commercial Landscape
商业景观

KEYWORDS 关键词

Plant Community
植物群落

Landscape Space
景观空间

Green Ecology
绿色生态

European Style
欧式风格

Location: Hedong District, Tianjin
Client: Tianjin Wanda Plaza Investment Co., Ltd.
Landscape Design: Palm Design Co., Ltd.
Palm Landscape Planning & Design Institute
Total Landscape Design Area: 64,868 m²

项目地点：天津市河东区
项目委托：天津万达中心投资有限公司
景观设计：棕榈园林股份有限公司
　　　　　棕榈景观规划设计院
总景观设计面积：64 868 m²

Tianjin Hedong Wanda Plaza
天津河东万达中心

FEATURES 项目亮点

The design needs to redivide the functional spaces and pay attention to people's activities and the landscape details. Starting from the magnificence of the landscape, the designers try to embody the exchange of experience and realize the functions within the site.

本案设计既要兼顾功能空间的重新划分，又要关注人的活动需求以及景观细节的体现。设计师从景观设计现象学意义上出发，力求体现该场地中人与人的经验交织，以实现超越物质和功能的需要。

Overview

Tianjin Wanda Plaza locates in Hedong District, covering a total landscape area of approximately 64,868 m², of which, about 62,870 m² is within the boundary lines. According to the client's requirement, the Plaza includes five parts: the commercial street, the luxury hotel, the roof garden atop the hotel building, the luxurious residences and the demonstration residential area.

项目概况

天津万达项目位于天津市河东区，总景观设计面积约 64 868 m²，其中红线范围内地面和用地红线至道牙部分景观设计总面积约 62 870 m²。按甲方要求，整个项目分成了商业街、酒店、酒店屋顶花园、豪宅以及豪宅示范区五个区域。

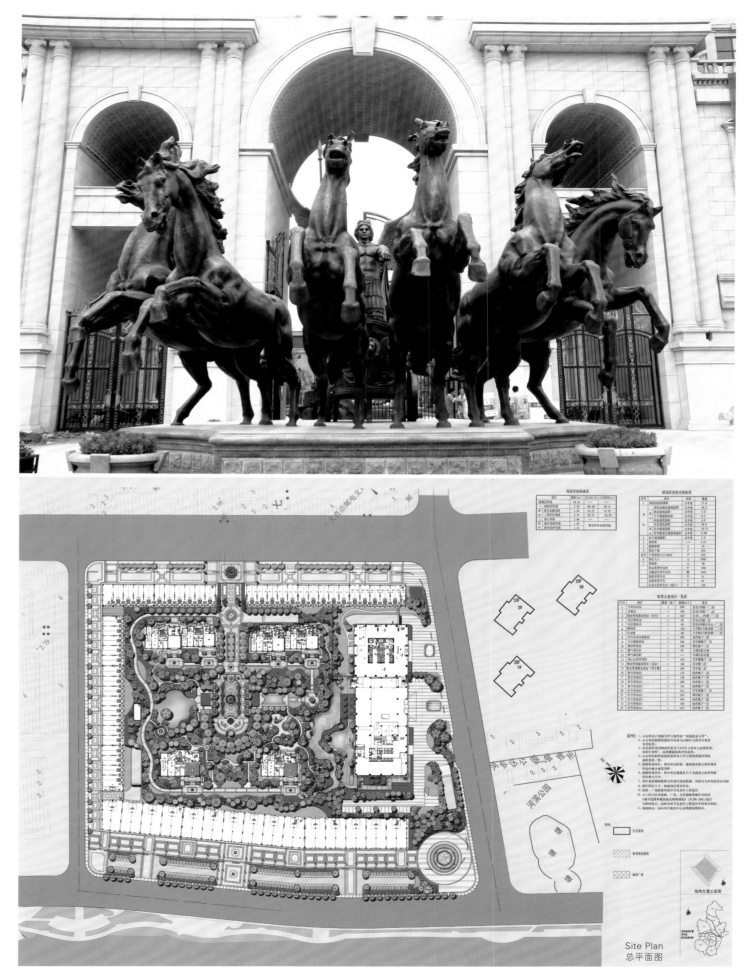

Site Plan
总平面图

Green Ecology

The functions and demands of the site have changed a lot. To create opulent landscape spaces, the easiest and most effective way is to create a multi-level plant community. Meanwhile, the functional spaces within the site need dense plants to cover the messy traces and provide enough shades. Thus, the existing trees such as the black pines are reserved, and local tree species such as Chinese scholar trees and apricot trees are added. All of these trees are well organized according to their growth characteristics to achieve quick succession.

Ever since the design began, the designers have decided the principles for construction. Tall trees are carefully transplanted with their canopies being well kept to gesture elegantly and adapt to the new environment quickly. Medium-rise flowering shrubs combine with the ground cover, grass and flowers, as well as the deciduous and evergreen trees to present different views for every corner at any time. Opulent plant community is established in a short period to revive the green ecology.

绿色生态

场地的定位和需求发生了巨大的变化。打造丰富的景观空间，最直接最有效的方法就是营造层次丰富的植物群落并自我演替。同时，场地各功能空间需要浓密的植物群落对场地进行二次划分，消隐零乱的痕迹，提供良好的遮阴。故对原有乔木进行选择性的保留（如黑松等），适当选用当地本土树种（如国槐、山杏等）进行合理搭配，尊重植物生长的生态法则，适地适树，促其演替。

设计伊始，设计师对施工工艺进行限定，采用大树全冠移植，保证其完整优美姿态的同时，又可尽快还原生态。中层花灌木结合地被、草花，落叶和常绿搭配，做到处处有花可观、时时有景可看。丰富的植物群落在短时间内完成地块性质的转变，促使绿色生态的重生。

Cultural Connotation of Landscape

Culture carries the memories and ideas of an ethnic group, and the cultural perception shows individual's cognition to the history of the group. It interacts directly with the world we live in. The project has made reference to European mythology to create a "Garden of Goddess" — "Ilisia". It will be the courtyard for goddess, a paradise on earth, and a beautiful place for living.

In the Garden of Goddess, there's the "River of Light" which symbolizes wisdom, victory and freedom; the "Crete Garden" — the most beautiful Aegean island; the "Luna Garden" which collects the most beautiful moonlight in the world. It is home to six goddesses: Goddess of Youth (Muse), Miss Liberty, Goddess of Victory (Nike), Goddess of Wisdom (Athene), Goddess of Dawn (Aurora) and Goddess of the Moon (Selene).

Zigzag paths and well-proportioned spaces have shown great gardening skills, while six goddesses in European mythology have created a graceful European-style garden.

景观文化内涵

　　文化是一个族群的记忆和观点,对文化的感知则是个体对族群生存史的体验,它与直接经验的世界相互交错影响。本案中借鉴欧洲神话,将本案打造成"众神之苑"——"伊利西亚",将这里变成众神的庭苑,人间的乐土,高贵而美丽的居所。

　　神的庭苑里,有"光之河",容纳智慧、胜利和自由的河流;有"克里特花园",爱琴海最美的岛屿;有"月神庭院",能够收集世界上最美丽的月光。这里居住着六位女神:青春女神缪斯、自由女神 Miss Liberty、胜利女神尼克、智慧女神雅典娜、黎明女神欧若拉、月亮女神狄安娜。

　　弯曲的园路和疏密有致的空间展示了造园的精髓,而欧式神话中的六位女神则把园区打造成了一个精美的欧式庭园。

Commercial Landscape
商业景观

KEYWORDS 关键词

Theme Landscape
主题景观

Bright and Colorful
绚丽多彩

Roof Garden
屋顶花园

Modernism Style
现代主义风格

Location: Fuzhou, Fujian
Developer: Beijing Thaihot Real Estate Development Co., Ltd.
Landscape Design: L&A Design Group
Landscape Design Area: 30,000 m²

项目地点：福建省福州市
发展商：北京泰禾房地产开发有限公司
景观设计：奥雅设计集团
景观设计面积：3万 m²

Thaihot City Plaza, Wusibei, Fuzhou
福州五四北泰禾城市广场

FEATURES 项目亮点

The five most valuable colored diamonds are introduced into the concept of commercial streets. By endowing five squares with different colors, it creates a "galaxy of floor lights" which shines at night to enhance the commercial atmosphere and increase fun for the site.

将钻石中最名贵的五种彩色钻石引入商业内街概念中，通过为五个广场赋予不同的颜色，形成夜晚五色变幻的"银河"地灯，增强商业氛围与场地乐趣。

Overview

The project is mainly composed of low-storey commerce, business office and centralized business, located in Xindian New Town Center of Jin'an District, Fuzhou City, which is one of the eight New Towns in Fuzhou overall planning publicity, containing relatively perfect supporting facilities.

项目概况

该项目位于福州市晋安区新店新城中心。新店新城是福州市总体规划公示中的八个新城之一，有较为完善的配套设施。项目产品主要由底层商业、商务办公和集中商业三部分组成。

Site Plan
总平面图

Landscape Design

The design introduces the concept of "diamond" into the whole project to correspond to each product: Soho is the mountain – 5 towers in a row like 5 mountains echo with the project's location and Fuzhou City's terrain features. Mall building is like the mined loose diamond attached itself to the peripheral site just like the rock pedestal in architectural form. Meanwhile, the five most valuable colored diamonds are introduced into the concept of commercial streets. By endowing five squares with different colors, it creates a "galaxy of floor lights" which shines at night to enhance the commercial atmosphere and increase fun for the site.

Roof garden adopts the "floating kaleidoscope" theme, corresponding to the diamond shape and color. Movable tree pools and flower beds, colorful build body and flowers create multi-angle, multi-level new vision and experience, featuring both enjoyable tropical garden and funny bar.

景观设计

　　设计以"钻石"为主题概念,并将概念引入整个项目,与各部分产品相对应:Soho 为山——五栋塔楼好像五座山峰连成一线,与项目地理位置和福州地形特征相呼应,MALL 周边场地为"岩石托",Mall 建筑本身为开采出的裸钻——建筑形态与未经雕琢的裸钻一样,附着在岩石上。同时,将钻石中最名贵的五种彩色钻石引入商业内街概念中,通过为五个广场赋予不同的颜色,形成夜晚五色变幻的"银河"地灯,增强商业氛围与场地乐趣。

　　项目的屋顶花园采用"漂浮的万花筒"的主题,万花筒与钻石形状色彩相对应。移动的树池及花池,绚丽的构筑体和花卉,创造出多角度、多层次的新视野和体验,兼具热带花园和酒吧的欣赏、玩乐功能。

Commercial Landscape
商业景观

KEYWORDS 关键词

Beauty of Nature
自然大美

Blend into Surroundings
融于环境

Landscape Node
景观节点

Southeast Asian Style
东南亚风情

Location: Lingshui Li Autonomous County, Hainan
Commissioned by: Vcanland Holding Group
Landscape Design: CSC Landscape
Size: 215,000 m²

项目地点：海南省陵水黎族自治县
项目委托：红磡集团
设计单位：深圳市赛瑞景观工程设计有限公司
项目规模：215 000 m²

Lingshui Sandalwood Resort Phase II
陵水红磡香水湾度假酒店二期

FEATURES 项目亮点

Designed mainly in Southeast Asian style, the landscape is blended with the romance and warmth of the coastal lifestyle, showing respect to nature by every detail and embodying the characteristics of the Southeast Asian landscapes.

整个酒店的景观设计以浓郁的东南亚风情为主，将浪漫温馨的海岸生活气息融入到景观设计中。大到空间打造，小到细节装饰，都体现了对自然的尊重，继承了自然、健康的东南亚风格景观特质。

Overview

Strategically located in Perfume Bay of Hongkan, Lingshui Sandalwood Resort is near the sea and backed by the mountain. Occupying a broad view of the South China Sea and sitting against the famous Diaoluoshan Primeval Forest, Perfume Bay, following Sanya Bay and Yalong Bay, has becomes a new coastal destination in the tropical area. Just steps away from the Boundary Island, the Perfume Bay even has its private entertainment island.

项目概况

陵水香水湾度假酒店，地处红磡香水湾，依山傍海，地理位置优越，是继三亚湾、亚龙湾之后最受世人瞩目的热带海景旅游新目的地。独特的地理优势不仅让香水湾坐拥南海广阔热带海景，更是背靠历史悠久的吊罗山原始森林；海南首席玩海乐园分界洲岛与红磡香水湾咫尺相遥，可称香水湾的私属娱乐岛屿。

Landscape Design

Curving shore line follows the mountainous topography, serving as the beautiful background for the elegant buildings. High-end residences such as the low-density apartments, seaview townhouses and single-family houses combine together to create a modern residential complex in the coastal area.

Designed mainly in Southeast Asian style, the landscape is blended with the romance and warmth of the coastal lifestyle, showing respect to nature by all details and embodying the characteristics of the Southeast Asian landscapes.

Naturally built pavilion, bridge and platform; birds' twitter, fragrance of flowers and sound of insects; the light blue ceramic tiles of the swimming pool that features strong breath of the tropical sea, all contribute to a beautiful and magnificent garden. Landscapes and buildings integrate into the surroundings to build a holiday resort in this modern city. It not only shows a kind of design style but also leads a kind of lifestyle.

景观设计

以迂回蜿蜒的海岸地形契合渐次叠落的天然山势,镶嵌优雅建筑于自然大美之中。超低密度璀璨布局瀚海公寓、揽海联排、旷海大院独栋等上层业态,创生出一个极富时代神韵和山海灵性的海岸建筑群落。

整个酒店的景观设计以浓郁的东南亚风情为主,将浪漫温馨的海岸生活气息融入到景观设计中。大到空间打造,小到细节装饰,都体现了对自然的尊重,继承了自然、健康的东南亚风格景观特质。

亭、桥、榭台的自然融入,鸟语花香虫鸣的景色,充满热带海洋气息的天蓝色瓷砖泳池,让庭园美轮美奂。景观与自然环境交融,自然中找景,景观中建造住宅,是现代都市中难得的闲适的度假风情酒店。风格粗犷自然、休闲浪漫的环境氛围,与其说是一种设计风格,不如说是一种生活态度。

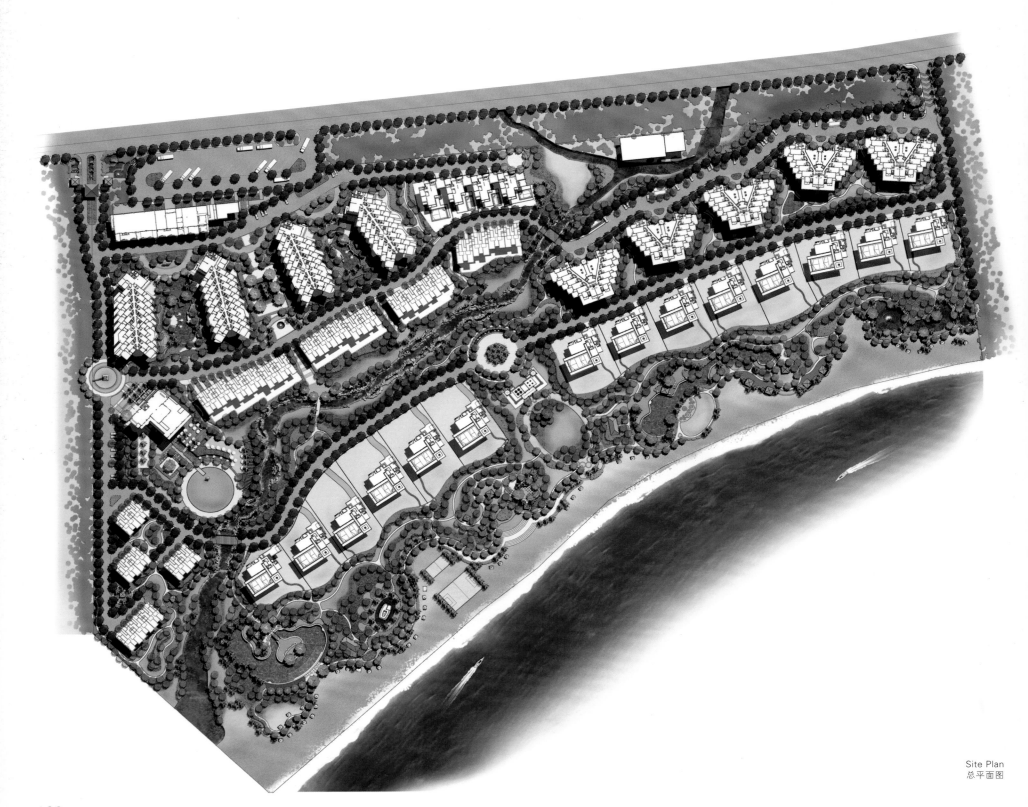

Site Plan
总平面图

Commercial Landscape
商业景观

KEYWORDS 关键词

Chinese Painting
中国画

Zen Ambiance
禅意氛围

Plant Space
植物空间

Modernist Style
现代主义风格

Location: Nanjing, Jiangsu
Developer: Yanlord Land Group Limited
Landscape Design: Botao Landscape (Australia)
Land Area: 15,000 m²

项目地点：江苏省南京市
开发商：仁恒置地
景观设计：澳大利亚·柏涛景观
占地面积：1.5 万 m²

Dream Island Sales Center, Nanjing
南京仁恒绿洲新岛销售中心

FEATURES 项目亮点

The project is located at Jiangxinzhou (a river island), which is an important connecting hub of Nanjing's cross-river developing strategy with a long-term planning to be built as an "Ecological Sci-tech city and Low-carbon Wisdom Island". The designers bring in this idea and combine the local cultural essence to build this featured and landmark landscape project.

项目所在地江心洲作为南京跨江发展战略的重要连接枢纽，其远景规划是建成一座"生态科技城，低碳智慧岛"。设计师引入这一理念，结合当地文化精髓，打造这个特色标志性景观设计项目。

Overview

Nanjing is near the tourist and cultural cities Yangzhou, Wuxi and Suzhou, which are all typical southern landscape areas, thus it's with superior location. The project is located at Jiangxinzhou (a river island), which is an important connecting hub of Nanjing's cross-river developing strategy with a long-term planning to be built as an "Ecological Sci-tech city and Low-carbon Wisdom Island". The designers bring in this idea and combine the local cultural essence to build this featured and landmark landscape project.

The project adopts the modernist style, which is full of modern sense. It's made into three plots as negotiation area, office space and showcase area. The building uses the combination of glass, grey stone, aluminum alloy and wood grill to create a transparent and clean character.

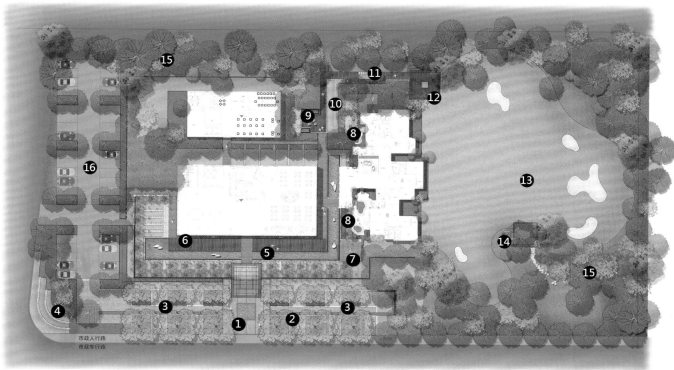

1. 主入口广场
2. 银杏树阵
3. 前场人行空间
4. 景观 LOGO 墙
5. 展示区水景
6. 亲水平台
7. 特色点景树
8. 精致庭院空间
9. 休闲平台
10. 特色草台阶
11. 趣味人行道
12. 观景廊架
13. 阳光草坡
14. 对景亭
15. 疏林草坡
16. 生态停车空间

Site Plan 总平面图

| LOGO指示区 | 转角特色节点 | 人行空间 | 主入口景观广场 | 人行空间 | 密林围合区 |

Front Plaza Area Elevation 前场区立面图

| 市政人行车行路 | 主入口前广场 | 特色入口空间 | 建筑 |

Exhibition Area Section 展示区剖面图

| 趣味休闲空间 | 特色草阶 | 特色节点 | 小径 | 主题景观亭 |

Leisure Area Section 休闲区剖面图

项目概况

南京紧挨着旅游文化城市扬州、无锡、苏州等典型的南方园林地带，有着十分优越的地理位置。项目所在地江心洲作为南京跨江发展战略的重要连接枢纽，其远景规划是建成一座"生态科技城，低碳智慧岛"。设计师引入这一理念，结合当地文化精髓，打造这个特色标志性景观设计项目。

项目建筑采用现代简约风格，极具时代感，共包含洽谈签约区，办公空间和样板展示区三大块。建筑采用玻璃、灰色石材、铝合金和木纹格栅相结合，打造出一种通透、干净的特质。

Landscape Design Objective

Based on the architectural design and the site situation, the designers get inspirations from the "island" and "rain". "The small island of Jiangxin floats on the boundless water, and when breezes come, the drizzles hit in the tranquil water surface and generate blossoming ripples. The small island is surrounded by drizzles and ripples, and the drizzles circle around, linger and generate ripples and sprays, adding some colors to the tranquil water surface. Let people in real life to experience this feeling, make the modern urban life with unlimited pursuit of efficiency and speed full of Zen ambiance, and let all things become slow and quiet." These are the designer's designing objectives of this project. By borrowing the expressing approach of Chinese painting, the designers strive to make the intention of life come out from the three-point perspective of still pictures: drizzle – jointly arranges with water to express the Zen sculpture with the meaning of raindrops, meanwhile borrows the cultural background of Nanjing's specialty riverstones to arrange the landscape stones; island – uses water island as a metaphor of water surrounded building; river surface – uses landscape water surface to surround the building and create Zen space.

景观设计目标

基于建筑设计和场地情况，设计师从"屿"和"雨"获得灵感。"江心的小岛漂浮在无边无际的水面上，一阵风吹来，细雨打在平静的水面上泛起一朵朵涟漪。小岛被细雨和涟漪环绕，细雨在此回旋、驻足、泛起涟漪、激起浪花，为平静的水面增添了几分色彩。让现实生活中的人们体验这种感受，让无限追求效率和速度的现代城市生活富有禅意，让一切慢下来，静下来。"这是本案设计师的设计目标。通过借用中国画的表达方式，设计师力求让生活的意向从三点透视的静止画面中走出来：细雨——与水结合设置表达雨滴之意的禅意雕塑，同时借用南京特产雨花石的文化背景布置景石；岛屿——运用水上岛屿比喻被水环抱的建筑；江面——运用景观水面环绕建筑，打造禅意空间。

Landscape Design

The front plaza space uses ginkgo tree arrays to set boundaries and avoid the disturbance of the surrounding environment, and to hint the transformation from dynamic space to static space. The ginkgo tree arrays on two sides highlight the plaza entrance in the middle; the designers combine the mirror surfaced waterscape to create featured paving plaza and use the soft furnishings and the night lighting effect to create the ceremonial sense of the entrance space. The showcase area uses the contrast approach, and the mirror surfaced waterscape combines with the leisure platform to form a large space and a contrast with the small enclosed space at the entrance of the porch, making the space open and close naturally and giving people a sense of being suddenly enlightened.

The sculpture in the water and the artistic chair on the platform create the overall Zen ambiance of the space. The showroom area adopts the approach of making the large space larger and the small space smaller, creates the delicate small space and echoes to the finishing touch effect of the showcase area. The designers build Japanese rock garden to create the Zen space, providing the interior of the showroom with a quite good viewing point. The leisure space as the terminal of the showcase area provides resting space, and passing through the fun road and the elevated corridor platform is the large green space. The delicate courtyard space as the resting point during the tour is arranged with interesting pedestrian space, and the elevated corridor platform is built at the end of the sight to attract people, and space for viewing sunshine lawn, open forest and grass slope are created at the back for people who stay. The designers create delicate plant area at the junction of the front plaza and the showroom, use the transition of road and space, arrange landscape trees at some parts and build large space sunshine lawn, open forest and grass slope, forming a contrast with the delicate small interior space and providing space for activities in the future at the same time.

景观设计

前场空间利用银杏树阵划分界限，屏蔽周围环境的干扰，暗示从动态空间进入静态空间。两边的银杏树阵突出中间的入口广场，结合镜面水景打造特色铺装广场，通过软装摆设以及夜间灯光效果营造入口空间的仪式感。展示区采用对比手法，镜面水景结合休闲平台形成大空间，与入口门廊封闭小空间形成对比，空间收放自如，给人眼前一亮、豁然开朗的感觉。

水中的雕塑和平台上的艺术座椅则营造整体空间的禅意氛围。样板区采用大空间做大、小空间做小的手法，打造精致小空间，与展示区呼应起到画龙点睛的效果。打造枯山水营造禅意空间，给样板间室内视线提供一个很好的观赏点。休闲空间作为展示区的终点提供休息空间，通过趣味道路及廊架平台抵达绿化大空间。精致庭院空间作为游览中途的停留点，设置趣味性人行空间，视线尽头打造廊架平台吸引人群，并在背景为停留者打造可观赏的阳光草坪与疏林草坡空间。前场与样板间连接处呼应的打造精致植物空间，通过道路空间转换，局部设置点景树，打造大空间阳光草坪疏林草坡，与内部精致小空间形成对比，同时可以为后期活动提供空间。

Riverside Landscape

滨江景观带

Ecological Corridor
生态走廊

Waterfront Landscape
滨水景观

Open Green Land
开放绿地

Riverside Landscape
滨江景观带

KEYWORDS 关键词

Public Green Land
公共绿地

Waterfront Platform
亲水平台

Large Grass Slope
大草坡

British Style
英式风格

Location: Huangpu District, Shanghai
Developer: Shanghai Waitanyuan Development Co., Ltd.
Shanghai Huangpu District Greenery and Public Sanitation Bureau
Landscape Design: The Coast Palisade Consulting Group (C.P.C. Group)
Size: 24,000 m²

项目地点：上海市黄埔区
开发商：上海外滩源发展有限公司
　　　　上海市黄浦区绿化和市容管理局
景观设计：加拿大 C.P.C. 建筑设计顾问有限公司
建设规模：24 000 m²

Landscape Design for the Bund Source 33 Park and Suzhou River Waterside Platform

外滩源 33 公共绿地及苏州河亲水平台环境景观设计

FEATURES 项目亮点

Based on the theme of "reproducing" and "rebuilding", the landscape design has fully explored the historical and cultural connotation of the city, carried forward the cultural context, and taken into account the functional requirements of the modern society, to make the Bund Source be a unique City Classic.

项目景观设计的基本出发点——以"重现"和"重塑"的设计主旨，充分挖掘城市的历史文化内涵，体现城市文脉的延续性，同时充分考虑现代社会的功能需求，将外滩源打造成为独一无二的城市经典。

Overview

It is the landscape design for the Bund Source 33 public green land and the Suzhou River waterside platform. No. 33 green land is the former site for the Garden of British Consulate, and the design needs to restore the garden and add some modern landscapes; the Suzhou River waterside platform is designed as a waterfront landscape belt for the city. Due to the nature of the land, the design should abide by the Bund's protection planning and Suzhou River's waterfront planning.

项目概况

项目景观设计包括两个方面的内容——外滩源33公共绿地和苏州河亲水平台。其中，33号公共绿地为原英国领事馆花园，将以风貌复原与现代景观设计相结合，而苏州河亲水平台则是作为城市滨水景观带来进行设计。其定位是开放的公共绿地和亲水平台，因而它必须符合外滩历史文化风貌保护规划和苏州河滨河景观规划。

Site Plan 总平面图

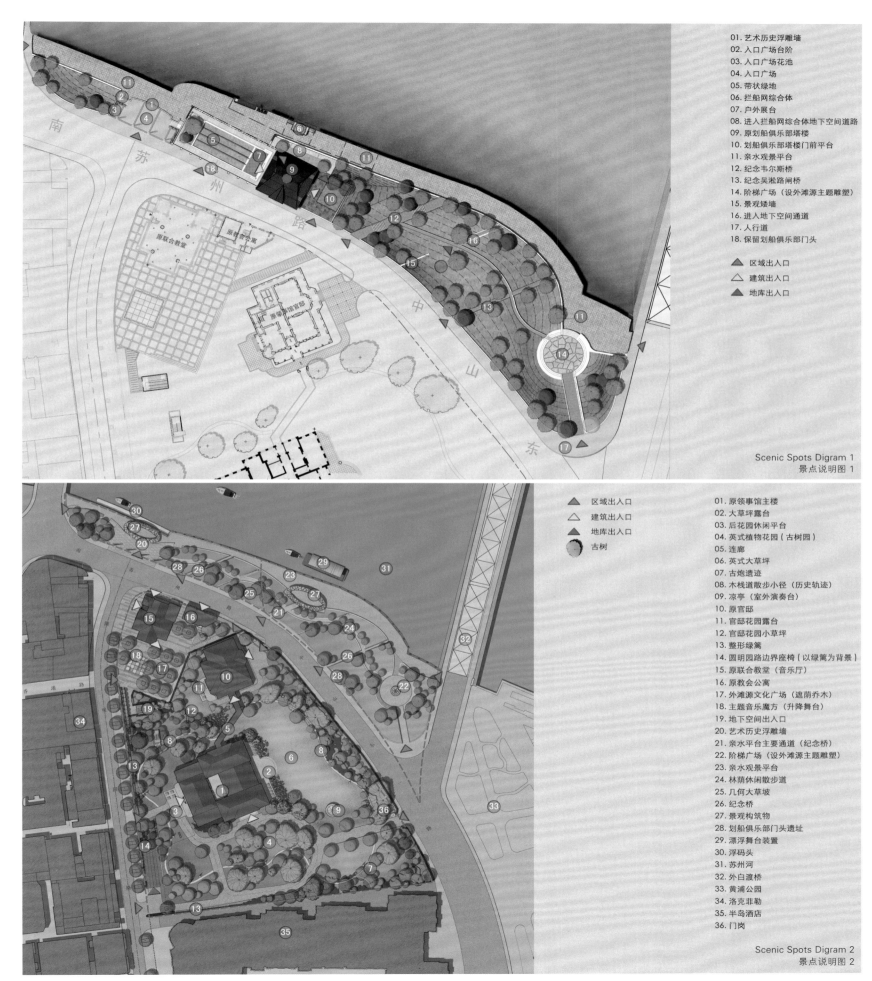

Scenic Spots Digram 1
景点说明图 1

Scenic Spots Digram 2
景点说明图 2

Design Theme

Based on the theme of "reproducing" and "rebuilding", the landscape design has fully explored the historical and cultural connotation of the city, carried forward the cultural context, and taken account of the functional requirements of the modern society, to make the Bund Source be a unique City Classic.

设计主旨

项目景观设计的基本出发点——以"重现"和"重塑"为设计主旨,充分挖掘城市的历史文化内涵,体现城市文脉的延续性,同时充分考虑现代社会的功能需求,将外滩源打造成为独一无二的城市经典。

Landscape Areas

According to different functions, the site is divided into three parts: Consulate Garden Area, Art Square Area and Waterside Platform Area.

1. Consulate Garden Area

The landscape design for this area has reserved and restored the typical characteristics of British gardens to transform the site into a glittering gem and present an elegant, tranquil and unpretentious garden of British Style. There are four key zones: parking entrance and botanical garden; big lawn and dinning terrace; Rose Garden; Mansion Garden.

The previous entrance is kept and changed to be a ring road for vehicles, which provides a simple route. Within the ring, a botanical garden with water features becomes the south garden for the main building.

As the living room for the main building, the big lawn is trimmed and surrounded by native plants. An outdoor dinning terrace is set at the exit to meet the requirements of the users. A pavilion which complements the garden can be used for wedding or photographing. Framed landscapes are created to draw people's attention.

Among the 27 old trees is a zigzag footpath that leads people to explore the value of the restored landscapes. Meanwhile, the old cannon and the monument base are kept as historical relics.

The Rose Garden nestles in the southeast corner of the site with a footpath running through. Planted with a variety of roses, the Rose Garden has brought bright colors for this green land.

The Mansion Garden sits on the southwest of the precious mansion with plants and lawns. A small outdoor terrace is set on the west side of the mansion, allowing people to enjoy the fresh air and have a cup of coffee.

2. Art Square Area

The art square locates in the northwest corner of the site, which is enclosed by the Union Church Buildings. It will not only be an outdoor performance space for the future concern hall, but also an ideal space for non-scheduled cultural and art performances in the Bund Source. The connection between the lawn of the mansion garden and the art square is designed with big steps for people to sit and watch the performances.

On the south of the art square, there's a sunken square that connects the underground commerce with the ground square. On the east, it is an access to the waterside platform, which also serves as the boundary between the art square and the mansion garden.

3. Waterside Platform Area

Taking advantage of the site's vertical character, it has transformed this riverfront green land into a big grass slope, thus created a three-dimensional plant space by artistry and design. The primary shorelines and the topographic texture are preserved as well as possible to keep the integrity of the historical context. At the same time, four famous bridges that have appeared in the history are abstracted and turned to be four channels crossing the big slope, to record the history and lead the city to the future.

景观分区

根据不同功能，分为三个区域：原领事馆花园区域、艺术广场区域和亲水平台区域。

1. 原领事馆花园区域

保留典型英式园林的特征是设计该区域景观的核心。它基本的特征将被复原、提炼和强化，就像一颗宝石重新回复它原有的光泽，呈现优雅、安静、低调的英式花园风格，有四个关键区域：入口车道与植物花园、大草坪与用餐露台、玫瑰园、官邸花园。

原来的入口在设计中被保留下来并形成一个车行环路，以提供一个更为便捷的循环路线，环路中间就强化形成了一个有水景的特色植物花园，形成主楼的南花园。

大草坪作为主楼的客厅，将加以整理，建筑周边种植基础植物。结合建筑东侧的出口设置露天用餐平台，更好的满足使用者的需要。一个凉亭的设置使花园的功能更加完整，可以用作婚礼或者摄影的

场所。透过浓密的植物修剪一些框景来引导人们的观景视线。在27棵古树之间设计了一条迂回的步行小径，这条小径将提供一条特殊的线索，来引导人们认识景观复原的价值。同时，古炮和纪念碑基座也作为一部分历史遗迹保留下来。

玫瑰园位于基地的东南角，是散步小径必经之地，栽满了各种玫瑰的花园将成为整个绿色浓荫中一抹明媚的颜色。

官邸花园位于原官邸楼的西南侧，以植物草坪为主，在官邸楼西侧设置了小型露天平台，可以享受清新的空气和美味的咖啡。

2. 艺术广场区域

艺术广场区域位于原领事馆基地的西北角，以原联合教堂建筑群为焦点形成的一个艺术广场空间，为未来的音乐厅提供了一个室外表演场所，也为外滩源地区不定期的艺术文化演出活动提供了一个适宜的空间。与领事馆花园的草坪连接处还设计了大台阶，为观众设置了一个可以停留观演的场所。

在此艺术广场南侧，为一个下沉式广场，是连接地下商业和地面广场的转换空间。艺术广场东侧设置了一条通往亲水平台的通道，也是与领事馆花园的分界。

3. 亲水平台区域

亲水平台区域利用了原有的竖向特征，使滨河绿地成为一个大地上的斜面大草坡，通过艺术化的手法创造一个更为立体的植物空间，使原地貌产生一种无形的张力，并尽量保持原有的岸线和地形结构不遭到破坏，保持历史信息的完整性和系统性。同时，设计者用抽象的手法把历史上出现过的四座桥梁演变为跨越大斜坡的四条历史通道，既记录了历史，也代表着城市从过去走向未来。

Riverside Landscape
滨江景观带

KEYWORDS 关键词

Cultural Landscape
人文景观

Urban Living Room
城市客厅

Cascades
跌水景观

Modern Simple Style
现代简约风格

Location: Zhangjiagang, Jiangsu
Developer: Zhangjiagang Municipal Government
Landscape Design: Botao Landscape (Australia)
Land Area: 65,000 m²

项目地点：江苏省张家港市
开发商：张家港市政府
景观设计：澳大利亚·柏涛景观
占地面积：65 000 m²

Zhangjiagang Xiaocheng River Renovation
张家港小城河改造

FEATURES 项目亮点

Defined as modern simple style, the landscape design of Xiaocheng River and Gudu Harbor covers the building facade design, the selection of stone materials and the landscape pavement, allowing people to experience the charm of the water town. Typical Jiangnan elements such as the pavilions and walls are used to interpret the lifestyle in the riverfront area.

立足于现代简约风格，小城河、谷渎港的改造从建筑立面的设计、石材选型、景观铺装，处处洋溢着浓郁的水乡风貌，以亭、榭及片墙元素，演绎具有江南特有的枕河人家。

Overview

Located to the south of the CBD's pedestrian street, Xiaocheng River is more than 2,200 m long and about 12 m wide, extending from Gudu Harbor in the east to Gangcheng Avenue in the west. Since the beginning of 1990s, half of the river has been covered by dwellings and been seriously polluted by rainwater and sewage. What's worse, the surrounding environment was disorderly, the traffic was crowded and the buildings were old and shabby.

项目概况

小城河位于城市核心商业区步行街的南侧，东起谷渎港、西至港城大道，全长2 200多米，河道平均宽度约12 m。上世纪90年代初期以来，由于近一半河道被房屋覆盖，雨污水直排河道，常年无法清淤疏浚，河道水质污染严重；小城河周边区域环境凌乱、交通拥堵、建筑破旧。

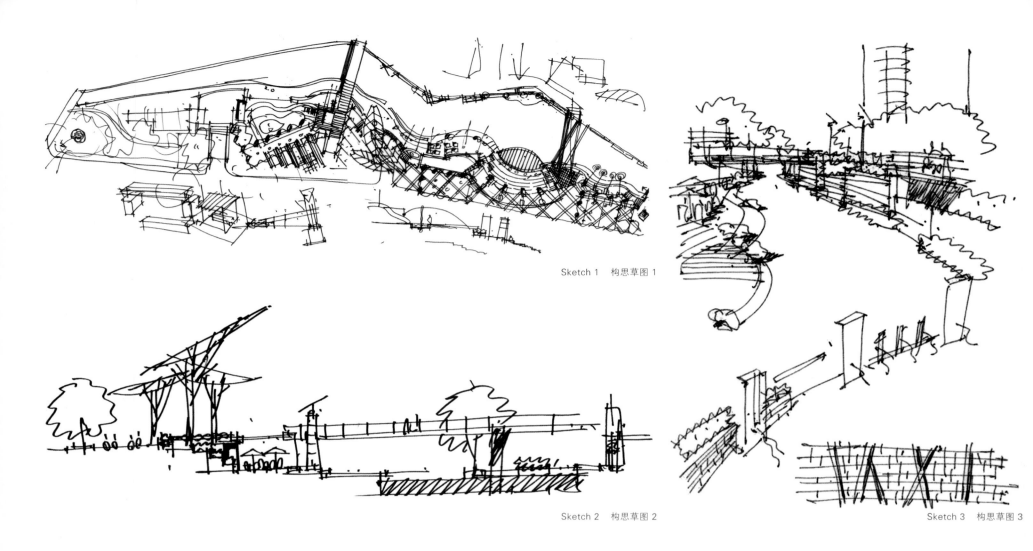

Sketch 1　构思草图 1

Sketch 2　构思草图 2

Sketch 3　构思草图 3

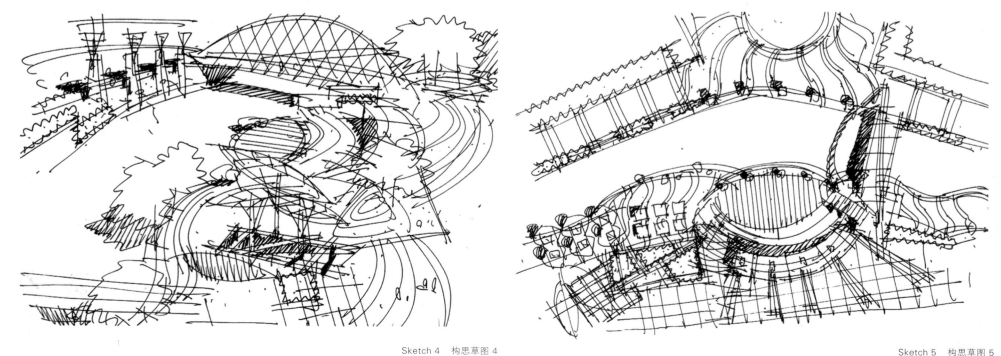

Sketch 4　构思草图 4

Sketch 5　构思草图 5

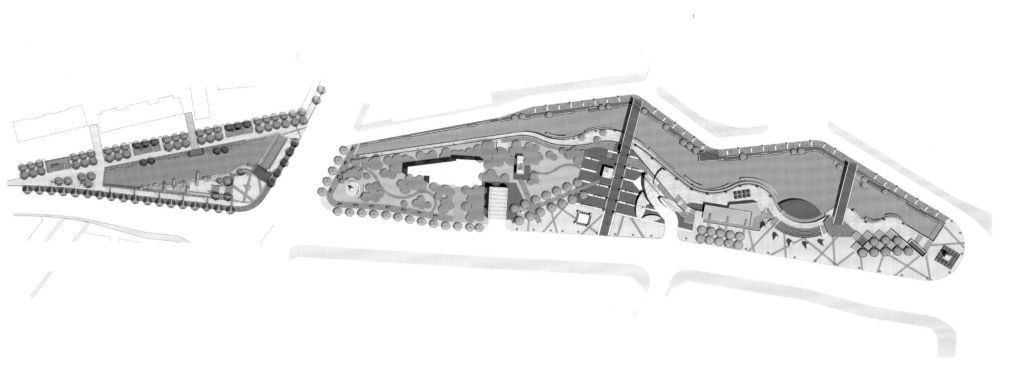

Overall Plan of Phase I
一期总平面图

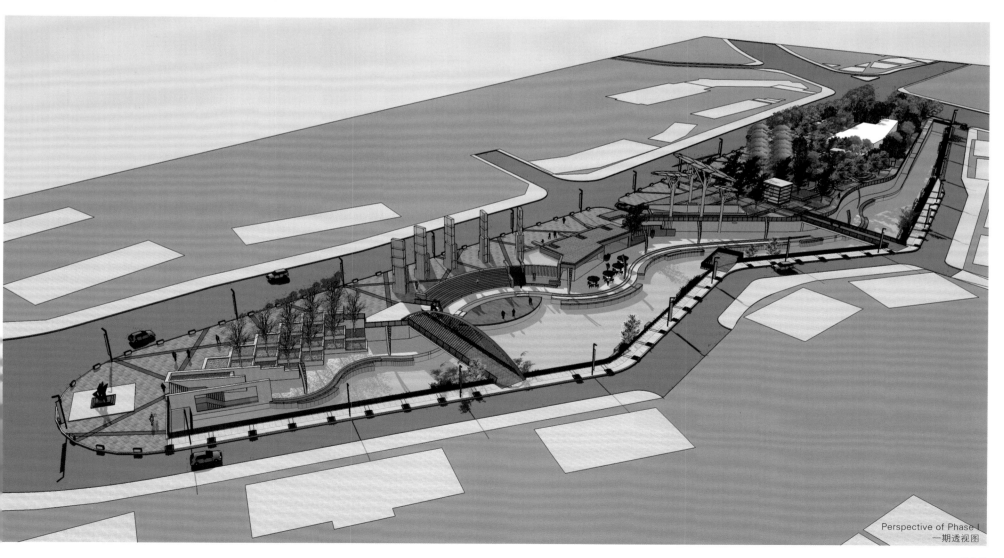

Perspective of Phase I
一期透视图

Overall Plan of Phase II
二期总平面图

Perspective of Phase II
二期透视图

Design Idea

Designed with the idea to create an "urban living room" in the CBD, the renovation project aims to restore the ecosystem of the river by sewage treatment and importing clean water. Through landscape design, it will provide a more comfortable environment and comprehensively upgrade the basic functions around this area.

Landscape Style

The landscape design of Xiaocheng River and Gudu Harbor is defined as modern simple style. From the design of building facade, the selection of stone materials and the landscape pavement, one can experience the charm of the water town. Pavilions and walls have been used to interpret the lifestyle in the riverfront. The designers also repaired the Qinglong Bridge that is familiar to the old Yangshe citizens. Thus, eight cultural elements in the old Yangshe's history, namely, Qinglong Bridge, Dragon's Howling, Gudu Tides, Lvxiang Pavilion, Bamboo Raft Wharf, Steamship Wharf, Jiyangmen City Wall and the Tablet with "Eight Forbiddances", are set along the river innovatively to show the cultural features of Gudu Harbor.

设计理念

小城河综合改造工程以打造核心商业区的"城市客厅"为理念，在治污引清的同时恢复河道自然生态，在建设景观的同时提供更加宜人的环境，在综合整治的同时全面提升整个区域的基础设施功能。

景观风格

立足于现代简约风格，小城河、谷渎港的改造从建筑立面的设计、石材选型、景观铺装，处处洋溢着浓郁的水乡风貌，以亭、榭及片墙元素，演绎具有江南特有的枕河人家。老杨舍人最熟悉的青龙桥得以修复，以青龙桥、龙吟、谷渎潮声、绿香亭、竹筏码头、轮船码头、暨阳门城墙及"八不准"碑等老杨舍历史上原有的八大文化元素，通过创意设置，点缀在谷渎港河道两侧，展现谷渎港滨水人文景观特色。

Central Landscape Belt

The center of the riverfront area is a place for citizen's activities. The landscape design here is modern and elegant to show the stability of the space by using slabstone and granite. The landscape bridge connects the riverbanks, providing easy access and also presenting beautiful views over the river. The central state is designed to enhance the lively atmosphere of the public square, while the riverfront footpaths will bring some romance to people. The cascades beside the coffee house drop from 3 m high, which will further enhance the atmosphere and bring a visual impact. The sculptures at the street corner will connect all these public spaces and form an integrated outdoor landscape system.

At the same time, the landscape design has fully respected the urban planning to well organize the bus station and provide great convenience for daily life. Both the landscape design and the architectural design have based on modern style while adding simple Chinese elements to realize the Chinese spirit.

中心景观带

滨河区域的中心地带，是广大市民的活动场所，景观设计大气、现代，通过条石、花岗岩的运用给人以厚实之感。景观桥是连接河道南北的中心纽带，方便广大市民的日常生活，同时也形成河道上的一道景观。中心舞台的设计增加了整个城市公共广场的热闹气氛，亲水步道的设计则为市民的生活增添了一份浪漫气息。咖啡馆建筑旁跌水的设计更加增加了广场的流动气氛，3 m的跌水落差给人以视觉上的冲击力，成为广场的一个景观点。道路街角对应雕塑的设计，使外部的公共空间连成一个主题，形成一个公共整体的室外景观。

同时，地块景观设计的时候尊重市政的规划设计，合理的安排了公共汽车站台，方便市民的日常生活。整个景观、建筑设计，中式精神大于中式元素，摒弃一切繁琐的符号，留下最安静的价值感。

219

Riverside Landscape
滨江景观带

KEYWORDS 关键词

Waterfront Platform
亲水平台

Environmental Art
环境艺术

Ecological Function
生态机能

Modernist Style
现代主义风格

Location: Guangzhou, Guang Dong
Landscape Design: Palm Design Co. Ltd.
Site Area: 510,000 m²

项目地点：广东省广州市
景观设计：棕榈设计有限公司
用地面积：510 000 m²

Foshan Dongping New Town Riverfront Landscape Zone
佛山东平新城滨河景观带

FEATURES 项目亮点

The scheme focuses on the protection and utilization of nature and has designed the largest wetland park in the city center of Foshan. And on the collocation of the plants, it focuses on the localism and restoration of the river's ecological function.

方案注重对自然的保护和利用，设计了佛山市最大的城市中心湿地公园，植物的配置强调乡土性，还原河流的生态机能。

Overview

The Foshan Dongping New Town Riverfront Landscape Zone was awarded the "Best Boutique" Prize in the first round of the annual city upgrading assessment of Foshan in 2014 by virtue of high-standard and high-quality greening landscape, and it has been recommended by the leaders of the city to various districts for learning and is also an important landscape image that the city promotes externally.

项目概况

佛山东平新城滨河景观带凭借高标准、高质量的绿化景观在 2014 年佛山市城市升级首轮年考中，获评"最佳精品奖"，多次被市领导推荐给各区学习，也是佛山市对外推介的一个重要景观面。

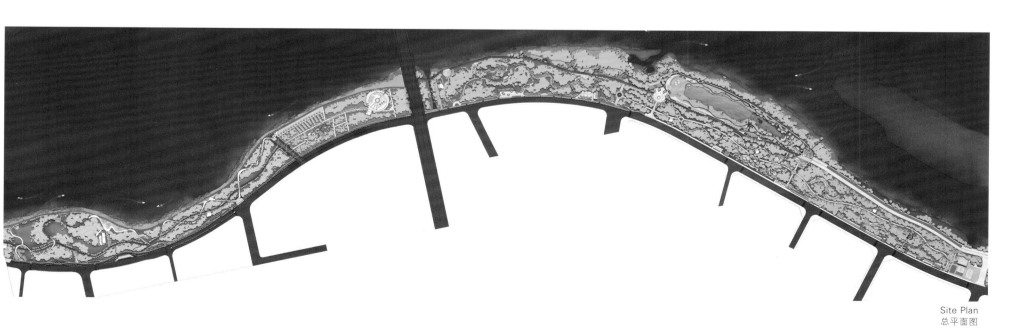

Site Plan
总平面图

图例 Legend

01 入口小平台	09 观演阶梯	17 入口特色构筑	
02 竹林幽径	10 表演舞台	18 特色健康步道	
03 树影特色平台	11 石湾陶瓷文化广场	19 龙舟主题雕塑	
04 林荫步道	12 休闲茶座	20 体育主题公园雕塑展示绿带	
05 球体雕塑群	13 现状休息平台	21 休闲构架	
06 阳光草坪	14 现状特色水景	22 文化展示小节点	
07 现状密林	15 木版门画文化展廊		
08 特色栈道构架	16 节点观景平台		

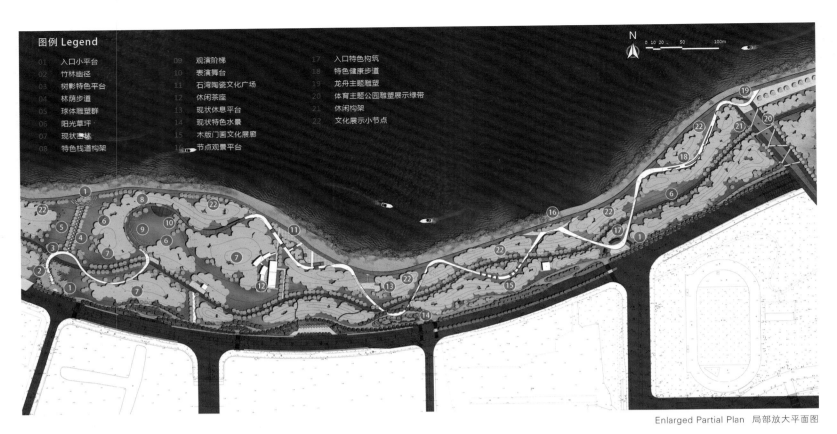

Enlarged Partial Plan　局部放大平面图

Landscape Design Idea

The project uses "water, green and fragrance" as design idea and uses "upgrading grade, increasing layer, adding color and culture" as design approaches, successfully integrating the grand landscape of Dongping River into the city. And it renovates the 8 km riverfront area into the city park with "distinct cultural theme and strong Lingnan characteristics", making it become the experience area where the general citizens enjoy the "high-efficiency and slow-paced life", and receiving consistent praise from the provincial and municipal visiting leaders at all levels as well as the general citizens.

景观设计理念

项目以"水绿香"为理念,以"升档次、加层次、增色彩、添文化"为手段,成功将东平河的胜景渗透进城市,把长约 8 km 的滨河片区改造成为"文化主题鲜明、岭南特色突出"的城市公园,使之成为广大市民享受"高效率、慢生活"的体验区,并且受到来访的省市各级领导以及广大市民的一致好评。

Landscape Design

City comes into being and becomes dynamic by virtue of water. Thus water becomes the inspiration of the design idea. During the flowing and ebbing of the tide, the water and the land are interweaved and permeated, and the strength of the tide has obscured the boundaries between the water and the land, making these two integrated. The design hopes to realize the dialogue between the city and the nature, and the communication between the modern life and the unsophisticated traditional culture through the organization of spaces and utilization of elements.

The diversified nature of the peripheral lands requires the function of the place to meet diversified demands of the comprehensive crowd that is considering the coordination between the function of the place and people's demands as well as paying attention to the display of the environmental art. The project adopts the measure of dividing the driveways and sidewalks, incorporates the municipal sidewalks into

景观设计

　　城依水而生，因水而律动。水成为设计思考的启动点。潮起潮落间，水与陆交织渗透，潮汐的力量使水陆界限模糊，二者在此达到互融。设计正是希望通过对空间的组织、对元素的运用，实现城市与自然的对话，实现摩登的现代生活与淳朴的传统文化之间的交流。

　　多样的周边用地性质要求场地功能满足综合人群的多样化使用诉求——既需要考虑场地功能与民生诉求的协调，又需要注重环境艺术的表达。项目采用人车分流的措施，将市政人行道纳入到公园慢性系统，规划了露天沙滩泳场、儿童乐园、滑板区等公众参与的功能区，同时将雕塑艺术融入场地设计，增添新城滨河景观

the chronic system of the park, plans public involved functional areas such as the open-air beach swimming pool, children's playground and skateboarding area, and integrates the statuary arts into the site design to enhance the cultural and artistic connotation of the New Town Riverfront Landscape.

The scheme focuses on the protection and utilization of nature and has designed the largest wetland park in the city center of Foshan. And on the collocation of the plants, it focuses on the localism and restoration of the river's ecological function. On the relationship between the water and the city, it combines with the city's developing elements; and the construction of the future wharf and beach swimming pool will make the distance between the city and the water closer; and getting close to water will become an important action of this city.

的文化艺术内涵。

方案注重对自然的保护和利用,设计了佛山市最大的城市中心湿地公园,植物的配置强调乡土性,还原河流的生态机能。在城水关系上,结合城市的发展脉搏,未来游船码头、沙滩泳池的兴建将拉近城与水的距离,亲水成为重要的城市行为。

Riverside Landscape
滨江景观带

KEYWORDS 关键词

Shoreline Landscape
岸线景观

Lively Space
活力空间

Landscape Node
景观节点

Modernist Style
现代主义风格

Location: Fuyang City, Zhejiang Province
Client: Zhongda Group/ Fuyang Zhongda Real Estate Co., Ltd.
Landscape Design: Palm Landscape Architecture Co., Ltd.
Chief Designer: Deng Daiming
Design Team: He Ziming, Kong Xiangfeng, Yu Weizeng, Liu Zhenxing, Yuan Yan, Tian Mengxiang, Huang Chengpeng, Qiu Jinhua, Liu Yuanzhi, Huang Xindong, Song Xuyang, Liao Zhi
Landscape Area: 87,000 m²

项目地点：浙江省富阳市
委托单位：中大集团／富阳中大房地产有限公司
景观设计：棕榈园林股份有限公司
主创设计师：邓代明
设计团队：贺子明、孔祥峰、余伟增、刘振兴、袁燕、田梦翔、黄成鹏、邱劲华、刘远智、黄新栋、宋旭阳、廖智
景观面积：87 000 m²

West Peninsula of Zhongda Group, Fuyang
富阳中大西郊半岛

FEATURES 项目亮点

The landscape planning obeys the design idea of fully discovering the landscape culture and folk art and culture of Fuyang city, strengthens and manifests it and endows it with the breath of times, and makes it integrated into the design of the waterfront landscape, aiming to create a waterfront lively space with unique charm.

景观规划秉承充分挖掘富阳的山水文化和民间艺术文化的设计理念，加以强调和演绎，并赋予时代气息，将其融入到滨水岸线景观的设计之中，旨在创造独具魅力的滨水活力空间。

Overview

The Fuyang West Peninsula waterfront project is located along the river segment between the Dapu Sluice and the foot of Daling Mountain, which is to the east of the downtown of Fuyang city and with a total length of about 2.1 km. It's with water in the front and mountain at its back, and various housing types are collocated along the waterfront shoreline, including hotel, commercial street, exhibition center, service-apartment, health club, SOHO, conference center, yacht, civil square and so on. According to the urban construction plan of Fuyang city, the West Peninsula of Zhongda Group will be built as a domestically famous and international first-class sports and leisure functional area that gathers competition organization, water sports, leisure and entertainment, exhibition and conference and living as a whole, and it's the "living room" of Fuyang city.

项目概况

富阳西郊半岛滨水岸线项目位于富阳市城区东侧的大浦闸—大岭山脚沿江段，全长约2.1 km，依山面水，沿滨水岸线配置各种各样的住宅类型，有酒店、商业街、会展中心、酒店式公寓、健康会所、SOHO、会议中心、游艇、市民广场等。根据富阳市城市建设规划，中大西郊半岛区块将建成为国内著名、国际一流的融合赛事组织、水上运动、休闲娱乐、会议会展、居住生活等功能于一体的运动休闲功能区，是富阳的"城市客厅"。

Landscape Design

The waterfront landscape planning strives to create an international community which is also rich with the local cultural characteristics of Fuyang city. The space and life should be internationalized with local traces. From the professional point of view, this is the first project that can truly integrate all the functions and create an urban life which can integrate with the landscape environment.

The landscape planning obeys the design idea of fully discovering the landscape culture and folk art and culture of Fuyang city, strengthens and manifests it and endows it with the breath of times, and makes it integrated into the design of the waterfront landscape, aiming to create a waterfront lively space with unique charm.

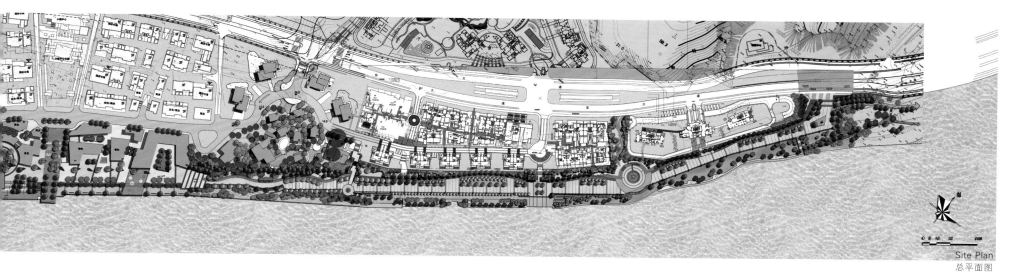

Site Plan
总平面图

景观设计

滨水景观规划着力创造了一个国际化的，但是又有富阳当地文化特征的社区。它的空间、它的生活是有国际化水准，同时又带有本地的痕迹。这个项目从专业的角度来讲，是第一个非常有特色的、把所有功能真正混合起来、能够创造一个城市生活又融入山和水环境的项目。

景观规划秉承充分挖掘富阳的山水文化和民间艺术文化的设计理念，加以强调和演绎，并赋予时代气息，将其融入到滨水岸线景观的设计之中，旨在创造独具魅力的滨水活力空间。

Landscape for Cultural & Civic Buildings

文化、市政建筑景观

Green Corridor
绿色廊道

Partitioned Greening
分区绿化

Landscape Node
景观节点

Landscape for Cultural & Civic Buildings 文化、市政建筑景观

KEYWORDS 关键词

Artistry
艺术性

Concise and Symmetrical
简洁对称

Central Waterscape
中心水景

Modern Chinese Style
现代中式风格

Location: Dongying, Shandong
Client: Dongying City Housing and Urban-Rural Development Committee
Landscape Design: Dongda Landscape Design
Size: 11,500 m²

项目地点：山东省东营市
委托单位：东营市住房和城乡建设委员会
景观设计：深圳市东大景观设计有限公司
项目规模：11 500 m²

Dongying City Lv Opera Museum Square Landscape

东营市吕剧博物馆广场景观

FEATURES 项目亮点

Landscape adopts new modern Chinese style, integrates the traditional culture elements with architecture and site by the modern design techniques and focuses on dramatic and artistic expression.

景观风格上，采用比较新的现代中式风格，将传统文化元素，通过现代的设计手法，融合与建筑与场地中，集中表现戏剧性和艺术性。

Overview

The Lv Opera, also known as "fancy dulcimer" and "Qin Opera", is popular in part area of Shandong, Jiangsu and Anhui, originated from delta of the Yellow River in the north of Shandong, and evolved from the Shandong Story-telling Opera, so far there are 100 years of history. Project site is located in the Yellow River Estuary cultural market, formerly known as the Dongying City Performing Arts Auditorium.

项目概况

吕剧，又名"化装扬琴""琴戏"，流行于山东和江苏、安徽部分地区，起源于山东以北黄河三角洲，由山东琴书演变而来，迄今有100年历史。项目基地位于东营黄河口文化市场内，前身为东营市艺术团演艺厅。

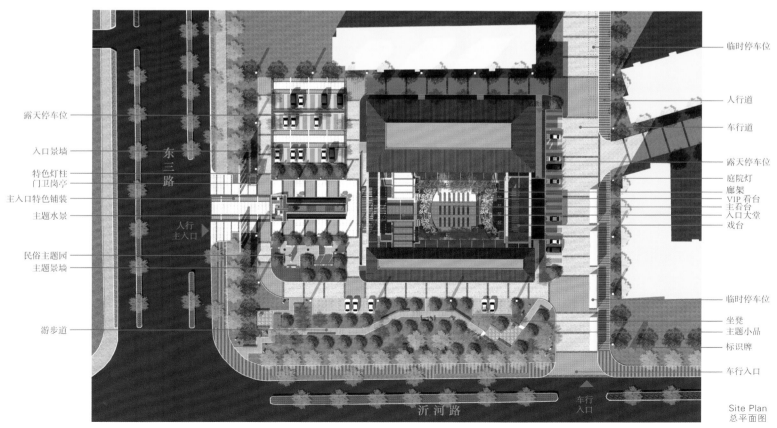

Site Plan
总平面图

Design Concept

The project design concept of Dongying Lv Opera culture plus modern Chinese style fully develops the Dongying Lv Opera's traditional culture, and it combines with modern Chinese style to create a landscape image of modern cultural museum.

Through the study and exploration of traditional museums, cultural centers and Quyi performance stage, the overall layout adopts central axis symmetry style and mainly uses the form of linear partition. Even the pavement is also linear, simple and symmetrical form.

Landscape Style

The landscape adopts new modern Chinese style, integrates the traditional culture elements with the architecture and the site by the modern design techniques and focuses on dramatic and artistic expression.

The reflection of culture is the focus of landscape expression. The entrance, the center water features, Quyi theme park, and center stage, all focus on the expression of the Lv Opera culture.

设计理念

项目以"东营吕剧文化＋现代中式风格"为设计理念，充分挖掘东营吕剧的传统文化，结合现代中式风格，营造出现代文化博物馆的景观形象。

通过对传统博物馆、文化馆及曲艺表演舞台的研究探索，整体布局上采用中心轴线对称的风格，形式多为直线划分、铺装亦是直线、简洁对称的形式。

景观风格

景观风格上，采用比较新的现代中式风格，将传统文化元素，通过现代的设计手法，融合与建筑与场地中，集中表现戏剧性和艺术性。

文化的体现是景观表达的重点，入口、中心水景、曲艺主题园、中心戏台，都有对吕剧文化集中体现。

Landscape for Cultural & Civic Buildings 文化、市政建筑景观

KEYWORDS 关键词

Green Ecology
绿色生态

Sci-tech Landscape
科技景观

Landscape Node
景观节点

Modernism Style
现代主义风格

Location: Luoyang, Henan
Developer: Luoyang Zhejiang Merchants Science and Technology Park Development Co., Ltd.
Architectural Design: Henan Zhibo Architectural Design Co., Ltd.
Planning/ Landscape Design: Shanghai Tianhe Runcheng Landscape and Planning Design Co., Ltd.
Green Coverage Rate: 35%

项目地点：河南省洛阳市
开发商：洛阳浙商科技园发展有限公司
建筑设计：河南智博建筑设计有限公司
规划/景观设计：上海天合润城景观规划设计有限公司
绿化率：35%

Science and Technology Innovation Park of Zhejiang University, Luoyang

洛阳浙大科技创意园

FEATURES 项目亮点

Four steps during ship's navigation are blended into the landscape design, highlighting the characteristics of each period and reflecting the struggle experience of outstanding entrepreneurs through their own efforts and teamwork.

在景观设计上把船舶航行的四个不同时期融入进项目设计中，重点突出每个区域的阶段特性，反映出创业者经过自身的努力和团队协作的精神成长为优秀企业家的奋斗历程。

Overview

The site is located on the south bank of the Luo River in the southwest of Luoyang City, to the east of Sunxin Road, to the south of Mudan Avenue and to the north of the Kaiyuan Road, with superior geographical position and convenient traffic. The project is in the radiation scope of Luolong High Tech Industrial Park, reaching east to Luoyang Higher Education Park, providing some population basis for the development. Project product types include corporate headquarters office, eco-friendly office and private apartment, developed in two phases.

Luolong district is a key high-tech industrial park in Luoyang City, and the information service industry is the leading industry to create the "Central Plains Sound Valley" and the industrial agglomeration area dominated by photovoltaic silicon, advanced equipment manufacturing, new energy and new materials. The government gathers and develops industries better to create a comfortable and beautiful living office environment in Luolong District.

Site Plan
总平面图

项目概况

基地位于洛阳市西南面洛河南岸，东临孙辛路、南接牡丹大道，北靠开元大道，地理位置优越，交通便捷。项目在洛龙高新技术产业园辐射范围内，东接洛阳高教园区，对本案开发提供了一定的人群基础。项目以企业总部办公、生态办公和人才公寓为产品类型，分二期开发。

洛龙区作为洛阳市重点高高新技术产业园区，以信息服务业为主导产业，打造"中原声谷"和硅光伏、先进装备制造、新能源新材料三大主导的产业集聚区。通过政府对产业的聚集和发展，更好得营造洛龙区舒适优美的人居办公环境。

Design Concept

The design starts from Luolong port and sets the ship's navigation process (starting – mooring point – preparing for sailing – exploration) as the streamline. These four steps are blended into landscape design, highlighting the characteristics of each period and reflecting the struggle experience of outstanding entrepreneurs through their own efforts and teamwork.

设计构思

本次构思以洛龙港为出发点，从船舶的起航—泊点—备航—探索为流线，在景观设计上把船舶航行的四个不同时期融入进项目设计中，重点突出每个区域的阶段特性，反映出创业者经过自身的努力和团队协作的精神成长为优秀企业家的奋斗历程。

Landscape Design

The project seizes the pulse of the times, by creating landscape, combining with the local industry characteristics, and mastering the stage characteristics of IT from information source – information transmission – information processing – information display – information storage, to build a full range of landscape in the park. At the same time getting close to the theme orientation of "the cradle of entrepreneurs, the harbor of winners", the project displays the entrepreneurs' life course from hard struggle to harvest and share after the success. Landscape organically combines the content of science and technology and the successful course of people who engage in science and technology, creating the Science and Technology Innovation Park's current landscape.

景观设计

洛阳浙大科技创意园抓住时代的脉搏，通过景观的营造，结合当地产业特色，抓住信息技术的阶段性特征从信息源—信息传输—信息处理—信息显示—信息存储来打造园区全方位的景观空间。同时紧扣"创业者的摇篮，成功者的港湾"主题定位，演绎企业家从辛苦创业奋斗到成功后的收获和分享的人生历程。景观把什么是科技，并把从事科技的人群、他们的成功历程两者有机结合，创造了现在的科技创意园景观。